Transitioning Towards a Knowledge Society

Julia Gremm · Julia Barth
Kaja J. Fietkiewicz · Wolfgang G. Stock

Transitioning Towards a Knowledge Society

Qatar as a Case Study

Springer

Julia Gremm
Heinrich Heine University Düsseldorf
Düsseldorf
Germany

Kaja J. Fietkiewicz
Heinrich Heine University Düsseldorf
Düsseldorf
Germany

Julia Barth
Heinrich Heine University Düsseldorf
Düsseldorf
Germany

Wolfgang G. Stock
Heinrich Heine University Düsseldorf
Düsseldorf
Germany

ISBN 978-3-319-71194-2 ISBN 978-3-319-71195-9 (eBook)
https://doi.org/10.1007/978-3-319-71195-9

Library of Congress Control Number: 2017958723

Printed on acid-free paper

This Springer imprint is published by Springer Nature
The registered company is Springer International Publishing AG
The registered company address is: Gewerbestrasse 11, 6330 Cham, Switzerland

Contents

Abbreviations

DECC	Doha Exhibition and Convention Center
FIFA	Fédération Internationale de Football Association
GCC	Gulf Cooperation Council
GDP	Gross Domestic Product
GERD	Gross Domestic Expenditure on Research and Development
GSDP	General Secretariat for Development and Planning (Doha, QA)
HDI	Human Development Index
HMC	Hamad Medical Corporation
ICT	Information and Communication Technology
IDI	ICT Development Index
K-12	Kindergarten to the 12th Grade
LNG	Liquefied Natural Gas
MENA	Middle East and North Africa
MIA	Museum of Islamic Art
MICE	Meetings, Incentives, Conferences and Exhibitions
NRI	Networked Readiness Index
PISA	Program for International Student Assessment
QBIC	Qatar Business Incubation Centre
QF	Qatar Foundation
QIA	Qatar Investment Authority
QNCC	Qatar National Convention Centre
QNDS	Qatar National Development Strategy
QNL	Qatar National Library
QNRF	Qatar National Research Fund
QNV	Qatar National Vision
QP	Qatar Petroleum
QR	Qatari Riyal
QSTP	Qatar Science and Technology Park
QU	Qatar University
R&D	Research and Development

STM	Science, Technology and Medicine
SWOT	Strengths, Weaknesses, Opportunities and Threats
TIMMS	Trends in International Mathematics and Science Study
UAE	United Arab Emirates
VPN	Virtual Private Network

Introduction

Once rapidly grown due to large oil and gas reserves, the governments of the Gulf Cooperation Council (GCC) states are nowadays aware of ending resources and declining demand of oil- and gas-burning countries due to electric mobility and enhanced ecological awareness. Additionally, new oil production methods as hydraulic fractioning, increasingly used, for instance, in the USA, intensify competition on the oil markets.

One GCC member, the state of Qatar—a region that is on the threshold of change these days, which conducts a rather unusual type of economy and whose population has to prove itself in its own country—makes great plans for the era after the oil. Its government has the task to bring the country and its people on the right track—a track that transforms Qatar into a knowledge-based society (QF, 2013[1]). To achieve this goal and to spread the nation's task, the previous Emir of Qatar, Sheikh Hamad bin Khalifa Al Thani, and his political supporters introduced the Qatar National Vision (QNV) 2030 in 2008.

> The National Vision defines broad future trends and reflects the aspirations, objectives and culture of the Qatari people. By shedding light on the future, the Vision illuminates the fundamental choices that are available to Qatari society. Simultaneously, it inspires Qatari people to develop a set of common goals related to their future (GSDP, 2008, p. 2[2]).

For the supplementary Qatar National Development Strategy the main goal is to build on knowledge:

> As Qatar's economy diversifies more from its reliance on gas and oil, success will increasingly depend on the ability to compete in a global knowledge economy. Educating and training Qataris to their full potential will be critical to continuing progress (GSDP, 2011, p. 122[3]).

---

[1]QF. (2013). *Qatar Foundation Strategic Plan 2013–2023*. Doha, QA: Qatar Foundation.

[2]GSDP. (2008). *Qatar National Vision 2030*. Doha, QA: General Secretariat for Development Planning.

[3]GSDP. (2011). *Qatar National Development Strategy 2011–2016*. Doha, QA: General Secretariat for Development Planning.

Our study tries to clarify to which extend the plans and strategies are implemented until now. Is the country on a good path to build up a knowledge society and to establish knowledge-based development?

What does "knowledge society" mean? And why did we choose Qatar as a case study?

These days, in many countries of the world, a transition into an "information society" or a "knowledge society" takes place (Webster, 2006[4]). An "information society" is technologically defined; it is based on information and communication technologies (ICT). A knowledge society is an information society as well; however, here knowledge will be available for everyone at any time and any place (Linde & Stock, 2011[5]).

Information societies as well as knowledge societies are varieties of a "network society" (Castells, 1996[6]): there are information networks (e.g. the Internet), and there exist knowledge networks (between knowledge workers or members of the "creative class"; Florida, 2005[7], 2012[8]). "Knowledge" includes both tacit knowledge (bound to persons) and explicit knowledge (bound to documents) (Barth et al., 2017[9]). As the body of knowledge is subject of change, lifelong learning is necessary.

Knowledge production, information dissemination as well as knowledge consumption became essential factors in knowledge-based economies and knowledge societies. Knowledge production generates scientific articles, technological patents, governmental information, everyday knowledge and—probably most important—graduates ready for knowledge-intensive jobs. Information dissemination (Stock & Stock, 2013[10]) happens digitally via ICT or physically, be it face-to-face or through physical documents, e.g., books. Knowledge consumption presupposes skilled people who are able to comprehend and understand all knowledge items they retrieved and are further able to transform found knowledge into meaningful actions. The integration of knowledge into products and services leads on to knowledge economy; the integration of knowledge into politics and administration results in knowledge-based development.

Anyway, in a knowledge economy or in knowledge-based development, knowledge itself is not an end product. No man can live from knowledge alone.

[4]Webster, F. (2006). *Theories of the Information Society*. Abingdon, UK: Routledge.

[5]Linde, F. & Stock, W. G. (2011). *Information Markets. A Strategic Guideline for the I-Commerce*. Berlin, Germany, New York, NY: De Gruyter Saur.

[6]Castells, M. (1996). *The Rise of the Network Society*. Malden, MA: Blackwell.

[7]Florida, R. L. (2005). *Cities and the Creative Class*. New York, NY, London, UK: Routledge.

[8]Florida, R. L. (2012). *The Rise of the Creative Class: Revisited*. New York, NY: Basic Books.

[9]Barth, J., Fietkiewicz, K. J., Gremm, J., Hartmann, S., Ilhan, A., Mainka, A., Meschede, C. & Stock, W. G. (2017). Informational urbanism. A conceptual framework of smart cities. In *Proceedings of the 50th Hawaii International Conference on System Sciences, January 4–7, 2017, Waikoloa Village* (pp. 2814–2823). Washington, DC: IEEE Computer Society.

[10]Stock, W. G., & Stock, M. (2013). *Handbook of Information Science*. Berlin, Germany, Boston, MA: De Gruyter Saur.

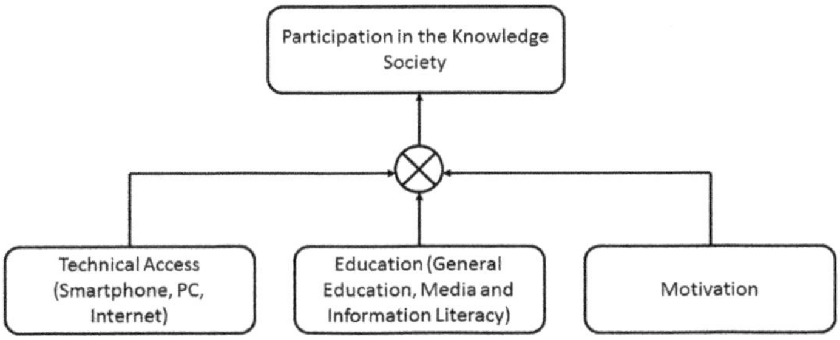

Fig. 1 Factors of successful participation in the knowledge society. *Source* (Linde & Stock, 2011, p. 95) (modified)

However, knowledge takes full effect if implemented in products and services, in entire markets and industries as well as in public governments. In such a way, "smart" products, "smart" services, "smart" industries, "smart" cities and "smart" countries may emerge. A necessary condition for the performance of a knowledge economy is the participation of a well-educated and motivated workforce.

What are the conditions for people to participate successfully in the knowledge society? Access to devices (such as smartphone, tablet, laptop or PC) and to the Internet as well as the possibility of financing these is a necessary condition. However, it only becomes a sufficient condition when it is joined by the motivations for dealing with the particularities of the knowledge society, and when levels of general education as well as of media and information literacy are reached that facilitates an adequate usage of digital media and knowledge sources (Fig. 1). So participation in the knowledge society does not hinge on technology and on knowledge and education alone. Additionally, it is important that the people really want to participate, thus there is adequate motivation (Linde & Stock, 2011, pp. 95 f.). In countries, where some citizens exhibit kinds of "rentier mentality" (Beblawi, 1987[11])—and Qatar is such a country—it could be problematic to change people's habits in terms of participation in the knowledge society or even in their willingness to work hard.

Based upon research on informational urbanism and informational (or "smart") cities (Stock, 2011[12], 2015[13]), a study on all major cities on the coast of the Arabian

[11]Beblawi, H. (1987). The rentier state in the Arab World. In H. Beblawi & G. Luciani (Eds.), *Nation, State and Integration in the Arab World. Vol. 2: The Rentier State* (pp. 49–62). London, UK: Croom Helm, New York, NY: Methuen.

[12]Stock, W. G. (2011). Informational cities: Analysis and construction of cities in the knowledge society. *Journal of the American Society for Information Science and Technology, 62*(5), 963–986.

[13]Stock, W. G. (2015). Informational urbanism. *Journal of Systemics, Cybernetics and Informatics, 13*(6), 62–69.

Fig. 2 Dusk at Qatar's Education City

Gulf (Gremm, Barth, & Stock, 2015[14]; Kosior, Barth, Gremm, Mainka & Stock, 2015[15]) determines Doha as the city with the highest potential to become an informational city, thus a prototypical city of the knowledge society. Based on these results, we decided to get to the bottom of it by bringing more detailed aspects into question and by understanding Qatar's aims and the nationals' way of thinking, living and working. Do they really have the potential to become a knowledge society as it is announced in the country's master plans?

In comparison to other oil-producing countries, Qatar and some other Arab states bank heavily on the development towards a knowledge society. When we take a look at Venezuela, which is one of the largest oil exporters on a global scale, we are not able to identify strategic plans towards knowledge economy or knowledge society. Many people in Venezuela live in poverty; the economy is in dire straits; there is social unrest. The dependency of Venezuela's economy on oil production and export, connected with rentier mentality of the political elite, led to maldevelopment and underdevelopment (Kingsbury, 2016[16]). In contrast to Venezuela, Qatar nowadays is a prospering and forward-looking state. When we visited Qatar, we found world-class infrastructures of knowledge institutions (Fig. 2).

---

[14]Gremm, J., Barth, J., & Stock, W. G. (2015). Kuwait is the past, Dubai is the present, Doha is the future: Informational cities on the Arabian Gulf. *International Journal of Knowledge Society Research, 6*(2), 51–64.

[15]Kosior, A., Barth, J., Gremm, J., Mainka, A. & Stock, W. G. (2015). Imported expertise in world-class knowledge infrastructures: The problematic development of knowledge cities in the Gulf region. *Journal of Information Science Theory and Practice, 3*(3), 17–44.

[16]Kingsbury, D. N. (2016). Oil's colonial residues: Geopolitics, identity and resistance in Venezuela. *Bulletin of Latin American Research, 35*(4), 423–436.

Another reason to select Qatar as a case study lies in the starting position for the transformation process towards a knowledge society. About two or three decades ago, in Qatar there was nothing we could associate with knowledge economy: no functioning education system K-12, nearly no universities, no knowledge-intensive companies, no knowledge-based development. In contrast, in "old" countries as the USA, the UK, France or Germany, knowledge institutions look back to centuries of history. In Qatar, we are able to study the transitioning towards a knowledge society from scratch, thus from the very beginning.

Qatar acts very successfully in terms of economic development, wealth of the native population and constructing the knowledge society—possibly even too successful. However, apart from envy, Qatar's position between two blocs of states culminated in the diplomatic crisis in 2017. On the one side, there is the Saudi-led Sunni bloc, on the other side the Iran-led Shiite bloc, Qatar is in the middle with relations to both. Due to the shared exploitation of the largest gas field of the world, Qatar and Iran have common interests and good relations in terms of business and foreign policy. There are also distinct economic and political relations as well as family ties between Qatar and its neighbouring countries, Saudi Arabia, UAE and Bahrain and the other GCC countries, Oman and Kuwait. The good relation between Qatar and Iran has annoyed Saudi Arabia for years. Furthermore, Qatar supported (and continues to partly support) foreign political parties as Egypt's Muslim Brotherhood and Palestine's Hamas both by financial aids as well as by media support through Qatar's Al Jazeera network. This foreign policy is not in line with Saudi Arabia's and emphasizes the special position of Qatar in the Middle East and North Africa (MENA). In June 2017, on the basis of fake news, some Islamic states (including Qatar's neighbouring states Saudi Arabia, UAE and Bahrain) cut off their diplomatic ties and isolated Qatar in the region. However, Qatar still maintains good political as well as economic relations to Japan, South Korea, India and China, to Turkey, Russia and to the USA as Qatar hosts the largest US military base in the entire Middle East region. As you can see, Qatar plays a particularly prominent role in the whole MENA region.

For Abdulwahed and Hasna (2017, p. 6[17]), there are four main drivers of Qatar's contemporary innovation- and knowledge-based economics. The first one is the Qatar National Vision, which is in close relationship to the second driver, namely the instability of oil and gas prizes. The mastering of the preparations for the Soccer World Cup (in 2022) and partly associated mega projects (including a new metro system, and with Lusail a new "smart" city) appears to be the third driver. Finally, the fourth driver is the market situation on the Arabian Peninsula with Saudi Arabia and the United Arab Emirates as Qatar's main competitors.

There are few other monographs on Qatar which appeared during the last years. "Demystifying Doha" by Salama and Wiedmann (2013[18]) is on Qatar's capital city,

---

[17]Abdulwahed, M., & Hasna, M. O. (2017). *Engineering and Technology Talent for Innovation and Knowledge-Based Economies*. Cham, Switzerland: Springer.

[18]Salama, A. M., & Wiedmann, F. (2013). *Demystifying Doha*. Farnham, UK: Ashgate.

its architecture and urbanism; however, Doha's position as a city of the knowledge society is not addressed. Kamrava's "Qatar. Small State, Big Politics" (2013; with a new preface 2015[19]) is a book about Qatar's politics, its international relations and its economy, but not on its knowledge society or knowledge economy.

Our book is arranged into four parts consisting of all in all 13 chapters. Part I (Chaps. 1 and 2) is on *Context*, namely the GCC states in general and Qatar as our case study. In the following Part II (*Concept*), we define in Chaps. 3 and 4 the basic concepts of knowledge society and introduce methods to operationalize the concepts. Part III (*Connection*) describes in Chaps. 5–12 the empirical research results of our studies on and in Qatar. Finally, Part IV with Chap. 13 is the *Conclusion* of our endeavours.

In Chap. 1, we will provide background information on informational ("smart") cities in the GCC states. To get an overview of Qatar, Chap. 2 gives basic information about the region and the country, the geographic and demographic data, the culture, the politics and the economy, the health care conditions and the education system. Chapter 3 will introduce the concepts of knowledge society and knowledge-based development and adds some further facts about Qatar by interpreting indicators of the development status of a country. Subsequently, the methods that underlie the study are described in Chap. 4. The above-mentioned strategies of the country as well as some others, like Qatar e-Government 2020 Strategy (ictQatar, 2014[20])—all related to the National Vision—drew attention to the topics discussed in Chaps. 5–12. In Chap. 5, we discuss different industries Qatar invests in to diversify its economy and to unlearn the reliance on oil and gas. The next chapter is about Qatar as an upcoming "inquiring, innovative and creative society" (GSDP, 2011, p. 11). The results of an e-Government study that analyses the government–citizen relationship are discussed in Chap. 7. Chapter 8 reviews the country's support of knowledge-based development. Chaps. 9–11 describe the higher education institutions and systems, its students and the graduates' ways into the labour markets. Chapter 12 completes—with an overview of the country's libraries, science parks and research funding—the information about the nation's knowledge-intensive institutions. A SWOT analysis in Chap. 13 will summarize all previous mentioned findings and points out Qatar's strengths, weaknesses, opportunities and threats on its way to a knowledge society.

One of the main research questions of our book is to identify problems and barriers of Qatar's way into the knowledge society. According to Qatar's master plans, this transformation will be finished in 2030. However, are those plans realistic? What is the state of the art of Qatar's institutions of higher education? Who does educate the students, Qataris or foreign talents? Do Qatar's people exhibit willingness, motivation and ability to create a knowledge society? Are there indeed jobs for highly educated native people or rather for expats? Based upon field

[19]Kamrava, M. (2015). *Qatar: Small State, Big Politics*. Ithaca, NY: Cornell University Press.

[20]ictQatar. (2014). *Qatar E-Government 2020 Strategy*. Doha, QA: Ministry of Information and Communication Technology.

research on-site, desktop studies and qualitative interviews with experts in Doha, we found that Qatar's way into knowledge society is by no means easy, but very problematic and arduous.

The book's target audience are economists, sociologists, political scientists, geographers, information scientists and other researchers on the knowledge society, policymakers in Qatar, but also all researchers and practitioners interested in the Arab oil states and their future.

Düsseldorf, Germany
November 2017

Julia Gremm
Julia Barth
Kaja J. Fietkiewicz
Wolfgang G. Stock

Part I
Context

Chapter 1
Informational Cities in the GCC States

Abstract Many cities in the world define themselves as 'smart'. Is this term appropriate for cities in the emergent Gulf region? This chapter investigates seven Gulf cities (Kuwait City, Manama, Doha, Abu Dhabi, Dubai, Sharjah, and Muscat) that have once grown rich due to large reserves of oil and gas. Now, with the threat of ending resources, governments focus on the development towards a knowledge society. We analyze the cities in terms of their cityness as well as their 'smartness' or 'informativeness' by a quantitative ServQual-like survey and by in-depth qualitative interviews (N = 34). Especially Doha in Qatar is well on its way towards an informational city, but also Dubai and Sharjah (both in the United Arab Emirates) make good scores.

Keywords Arab cities · GCC countries · Gulf cities · Informational city
Smart city · Knowledge society · Kuwait city · Manama · Doha
Abu Dhabi · Dubai · Sharjah · Muscat · Kuwait · Bahrain · Qatar
United Arab Emirates (UAE) · Oman

1.1 The Gulf Region

In the heat of the Gulf region 50 years ago, the desert dominates the landscape, the beaches are almost empty, and the few people living there work as pearl divers, fishermen, traders or peasants. Replaced by glittering facades, high-end hotels, artificial islands, huge shopping malls, and the tallest constructions on earth, the region nowadays attracts people from all over the world. Sources of wealth in the GCC countries have been huge resources of oil and natural gas (Fig. 1.1). The economically advanced nations in the Gulf region that outpaced most of the other Arab countries are Bahrain, Kuwait, Oman, Qatar, Saudi Arabia and the United Arab Emirates (UAE). These nations merged to an affiliation called Gulf Cooperation Council in 1981 (GCCSG, 2012). This part of the world with the fastest-growing population and a high, but still growing, Gross Domestic Product (GDP) faces a huge challenge today and in the following decades. The management

J. Gremm et al., *Transitioning Towards a Knowledge Society*,
https://doi.org/10.1007/978-3-319-71195-9_1

Fig. 1.1 Source of wealth in the GCC Countries: oil and gas

of energy, water and food resources to diversify the region into a non-hydrocarbon industry and the education and employment of the untrained young Arabian population need to be improved to ensure high living standards and sustainable growth in the future (EIU, 2009, 2010). With the goal to become competitive in the 21st century, the GCC states "are articulating new visions and implementing strategies and programs to propel their countries toward diversified and sustained knowledge-based economies" (Tadros, 2015, p. 4). This goal can just be achieved by ensuring progress in ICT and the education infrastructures of the countries, but especially by focusing more on human capital and the business environment (Kumar & van Welsum, 2013). "In the 2000s, the Gulf States all embraced the concept of 'knowledge economy' as integral to their ambitious programmes of economic diversification and development" (Ulrichsen, 2016, p. 77).

We are going to look behind the glimmering facades of the Gulf cities and describe and analyze their status as informational cities or smart cities, i.e. as cities of the knowledge society. This will be done by applying a customized questionnaire and interviews concerning questions about aspects of cities of the knowledge society.

In the GCC states, nearly all citizens live in a city—outside the cities, there is desert; in all states and emirates one city dominates the state or emirate. Additionally, on the Gulf coast cities hold the image of wealth and hope of a brilliant future. "The development of a post-oil economy in the Gulf has led to a

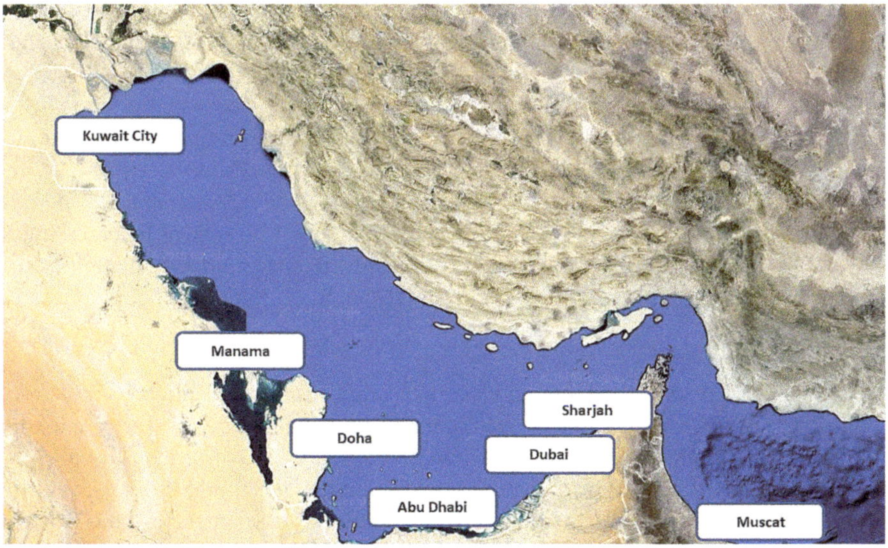

Fig. 1.2 Analyzed cities. *Sources* Kosior et al. (2015); Google Maps

new way of understanding cities as future asset that can sustain and widen the economic prosperity of the region" (Salama & Wiedmann, 2013, p. 56).

We will present results of our studies on-site (i.e., we visited all countries) and of further desktop research on all larger Gulf cities of the GCC states: Kuwait City (Kuwait), Manama (Bahrain), Doha (Qatar), Abu Dhabi, Dubai, Sharjah (United Arab Emirates), and Muscat (Oman) (Fig. 1.2). Apart from Greater Dammam (the center of Saudi Arabia's oil industry), there are no Saudi Arabian metropolises at the Gulf Coast; hence we neglected to study cities in Saudi Arabia.

Early on, the Kuwaiti government realized that they had to search for an oil alternative, and with it the Kuwait Vision 2035 arose: Kuwait shall be transformed into a trade and financial center, and invest in infrastructure projects such as a bridge between Shuwaikh in the north of Kuwait and the port village of Doha, a metro project and a future city. This major project of Kuwait named Silk City, scheduled to be completed in 2023, is an urban area which will be divided into separate districts: business, culture, and leisure. There are large numbers of financial services in Kuwait.

Bahrain, one of the first states profiting from oil, has focused its post-oil strategy on tourism and finance. The capital Manama has become the central place of realizing projects like new holiday resorts, modern shopping malls, and tourism offers and facilities. In 2012, Manama was voted to be the Arab Capital of Culture, indicating the traditional and cultural background of the city. Due to liberal regulations on alcohol trade and consumption which are unique in the Gulf Region, Manama is well attended for parties especially by people from Saudi Arabia. The offshore banking sector located in Bahrain's new Financial Harbor transforms

Manama into one of the most important financial service centers of the Middle East. Outside of Manama you can find the University of Bahrain and a Formula One circuit.

Since the late 1990s Qatar's capital Doha has developed rapidly into a city characterized by skyscrapers, shopping malls, museums, stadiums, and artificial islands. The Pearl, an artificial island with new apartments, houses, villas, a yacht harbor, parks, restaurants, and high-class retail strips, is already inhabited to a large part. The gorgeous waterfront (corniche) and the famous Museum of Islamic Art, as well as the fact that the city will be the venue of the soccer world championship, have put Doha into the limelight. Qatar has invested huge amounts of capital into its education system. Remarkable institutions are Qatar Foundation's Education City and Qatar University.

Abu Dhabi presents striking, modern skyscrapers along an almost eight kilometer long corniche. Tourism has been identified as an economic factor: the luxury hotel Emirates Palace, the Formula One speedway Yas Marina Circuit, the amusement park Ferrari World, and Saadiyat Island as the future culture center of Abu Dhabi are only some examples of the newest investments. Abu Dhabi's international airport carries about 16.5 million passengers annually and is the hub airport of Etihad Airways. Masdar City, an urban development project started in 2008 outside of the city will be prospectively one of the most sustainable communities for technology in the world and an emerging global hub for knowledge, research, business, and development. In proximity to Masdar City and the airport, Zayed University is located.

In recent years Dubai has gained attention from investors and the city's awareness increases steadily. As a consequence Dubai is developing into an important location in the range of trade, logistics, tourism, and services. Companies from all over the world settle down there. Tourist attractions like one of the most luxurious and expensive hotels (Burj Al Arab), the tallest building in the world (Burj Al Khalifa), and the artificial island The Palm draw millions of people from all over the world. Thereby Dubai has become a constantly developing global metropolis with an international airport which carries more than 66 million passengers a year. The airport is the hub for Dubai's international airline Emirates. There are several free trade zones, called "cities" (such as Internet City, Media City, International Academic City, etc.), hosting mostly foreign owned knowledge-rich institutions and companies.

Sharjah, the third-largest emirate of the UAE in terms of size and inhabitants, tries on one hand to maintain its conservative Islamic heritage, receiving the title Capital of Islamic Culture in 2014 by UNESCO, and on the other hand to adopt measures to realize the transition into a modern society. The steady development of Sharjah's economy and infrastructure, the expansion of the economic free zones, and also the emirate's ideal position right in the middle of the UAE, bordering all of the other six emirates, the Arabian Gulf and the Gulf of Oman, engages the interest of many investors. Just outside of Sharjah, a large campus provides the infrastructure for the University of Sharjah and the American University of Sharjah.

Oman's capital and seat of government is Muscat Capital Area, referred to as Old Muscat before 1970. Muscat has expanded its urban area in recent years. New projects are arising in terms of real estate and tourism including The Wave, a venture consisting of residential homes, hotels, retail, leisure, a marina, and a golf course, and the Oman Convention & Exhibition Centre. Ten kilometers from Old Muscat the large campus of Sultan Qaboos University is located in a development area.

To stay neutral in the discussion about the right term for the body of water between Iran and Arabia (*Arabian Gulf* for the Arabians versus *Persian Gulf* for the Iranians) we will call it the *Gulf* in this book. Hence, every country on the Arabian Peninsula and those who abut on the Gulf counts as a country of "the Gulf region".

1.2 Background: Cities in the Knowledge Society

Nowadays, more than half of the world's population is living in cities. The world is still changing into an "urban world" (Kourtit, Nijkamp, & de Noronha, 2015). This is the basic idea of our research program: Just as there were prototypical cities of the industrial society (for example, Manchester in middle of the 19th century) or the service society (such as New York's Manhattan in the second half of the 20th century), there are (or will be in the near future) typical cities of the knowledge society.

Sometimes, those cities are labeled as "smart". Additionally, there are further approaches conceptualizing cities and regions of the knowledge society: The terms "ubiquitous city" (Shin, 2009) and "digital city" (Cocchia, 2014) refer to the city's well developed information and communication technology (ICT) infrastructures; research on the "knowledge city" concentrates on knowledge-based urban development and its institutions as higher education, science parks and libraries (Carrillo, Yigitcanlar, García, & Lönnqvist, 2014; Madanipour, 2011), and, finally, "creative city" studies the "creative class" (Florida, 2005) as human capital in urban spaces. All mentioned approaches address in each case important topics of the description, analysis and planning of urban regions in contemporary and future cities of the knowledge era, but they do not reflect the whole story.

In contrast, in our interdisciplinary and holistic framework we study all aspects of knowledge and information, be it digital or physical, man or machine generated, which has implications for cities, their spaces, their institutions and—most important—their people. Albeit ICT is the heart of a smart city, we may not forget knowledge both in the form of tacit knowledge (bound to persons) as well as in the form of explicit knowledge (bound to documents; Stock & Stock, 2013; Negre & Rosenthal-Sabroux, 2015). Following Castells (1989), we call such prototypical urban regions of the knowledge society "informational cities" (Stock, 2011) or "smart cities" (in the broader sense of the concept; Fietkiewicz & Stock, 2015). Informational cities are studied by "informational urbanism" (Stock, 2015;

Barth et al., 2017). This term was coined by Stallmeyer (2009) to analyze "spatial transformations brought about by informational developments" (Stallmeyer, 2011, p. 2).

What is a "smart" or an "informational" city? In some scientific studies, definitions or approaches of a "smart city" are collected (e.g., Albino, Berardi, & Dangelico, 2015; Chourabi et al., 2012; Cocchia, 2014; Nam & Pardo, 2011a, b; Neirotti, De Marco, Cagliano, Mangano, & Scorrano, 2014). All definitions stress the importance of ICT for the 21st century city (Caragliu, Del Bo, & Nijkamp, 2011). Smart cities bank on ambient intelligence in information spaces city-wide (O'Grady & O'Hare, 2012) as well as in homes (Cook, 2012). However, the term "smart city" is fuzzy (Cocchia, 2014; Hollands, 2008; Negre & Rosenthal-Sabroux, 2015) and has its origins rather in marketing than in science; for example, the computer company IBM holds the trademark "smarter cities" and applies it in its smarter city campaign (Søderström, Paasche, & Klauser, 2014). We can identify two different concepts of the "smartness" of cities, one in a narrower sense of the concept and the other in a broader sense.

- *Narrower concept*: In the "vision of a smart city", Hall et al. (2000) introduce urban centers of the future, which "secure environmentally green". Here, a smart city is "forward-looking on the environmental front" (Chourabi et al., 2012). This narrow concept of "smartness" is strongly linked to natural resources and energy, transport and mobility, and living conditions, in short, to the sustainable green and livable city.
- *Broader concept*: While Chourabi et al. (2012) emphasize a smart city being "an icon of a sustainable and livable city", they define "smartness" with eight critical factors (management and organization, technology, governance, policy context, people and communities, economy, built infrastructure, and natural environment) much broader. Similarly, Giffinger et al. (2007) define smart cities in a broader sense by an enumeration of essential characteristics, i.e. smart economy, people, governance, mobility, environment and living. Such a broad view of smart cities is anticipated by Castells as early as 1989. For Castells, such cities are prototypical for the network society (Castells, 1996, 1997, 1998).

Both concepts of "smart cities" are important for informational urbanism, but we always have to distinguish between the narrower concept (emphasizing the green and sustainable city) and the broader concept (stressing the entire city as "informational"). In this book, we use "informational city" and "smart city" (in the broader sense) synonymously.

Informational urbanism has theoretical background both in urban studies, city sociology, city economics, architecture and city planning on the one hand as well as in computer science and information science on the other hand. Both building blocks are held together by system science. The common object of all mentioned disciplines is the multi-loop control system "city" and its various subsystems of infrastructures, economics, politics and administration, spaces, as well as location factors; and additionally, the information behavior of its stakeholders (i.e. residents, companies, administrations, visitors, etc.).

A similar approach to analyze cities in terms of their informativeness is "urban informatics". While urban informatics emphasizes the roles of people, place and information and communication technology with a focus on cities (Foth, Choi, & Satchell, 2011), informational urbanism includes not only ICT, but all kinds of information and (tacit as well as explicit) knowledge.

How can we understand cities in the emerging knowledge society? Max Weber's "The City" (1921) is one of the first sociological analyses of urban regions in which he stresses the anonymity in cities in contrast to rural regions, their roles as capitalist market centers and as corporate bodies with a definable area. In contrast to Weber, for Georg Simmel, in 1903, the functional magnitude of a metropolis goes beyond its actual boundaries. The metropolis has impact on the mental life of its residents; insofar there is an intensification of emotional and nervous life (Simmel, 1903) due to information overload in cities. With Simmel's and Weber's works, city sociology was born. Regarding the information overload in metropoles and—following Simmel—its influences on the minds of residents, a rudimentary form of informational urbanism emerged as well.

From the 1960s onwards, the exploration of world cities has experienced a big boost. A remarkable example is "The World Cities" by Hall (1966), who defines world cities as centers of politics, commerce, finance, professional services, luxury consumption and entertainment. For Hall, administrative boundaries are not important; he considers connected regions (such as Randstad Holland—a city conglomerate in The Netherlands including Amsterdam, The Hague, Rotterdam, Utrecht and some smaller towns) as one world city. The economy of cities is analyzed by Jacobs (1969). Jacobs underlines the essential roles of human capital and competition between firms as externalities of economic success of cities. Manuel Castells' book "Informational Cities" (1989) exerts massive influence on city research. In the network society, as in informational cities, two spaces exist side by side: geographical space ("space of places") and the space of information, money and power streams ("space of flows") created via digital networks. In informational cities the space of flows outperforms the space of places.

Of course, pure "cities of bits" (Mitchell, 1995), "e-cities" (Mitchell, 1999) or "cyborg-selves in the networked city" (Mitchell, 2003) as introduced by William J. Mitchell are not possible. However, besides the physical infrastructures (like streets, rails or airports) with the digital ones a second set of infrastructures arises. Mitchell, in his "e-topia" (1999), describes the interplays between both types of infrastructures. So people act—driven by ICT and the Internet—not only in the physical space, but also in the area of information flows, i.e. in the digital space.

John Friedmann proposes the "world city hypothesis" (1986). World cities are the centers of the world economy and urban areas with dense patterns of interactions. In a second attempt (Friedmann, 1995), he defines world cities—following Castells—through their position in the space of flows. On top of Friedmann's world city hierarchy, we find London, New York and Tokyo. For Sassen (2001), "global cities" are centers of the globalized world. They host headquarters of major companies and their suppliers, first of all service providers (financial services, advertising, media, consulting, etc.). All these companies depend on information;

face-to-face information flows are as important as global digital information streams. The global city is an extremely intense and dense information loop, so that the city itself becomes an information center. Taylor (2004) shows that the world cities generate a network formed by leading service companies' power and information flows. Taylor also stressed the "green" component of cities that makes these worth living in. Taylor, Hoyler, and Verbruggen (2010) use the term "cityness" to describe the position of a city in the global network (as opposed to the "townness", which captures the role of a city to its hinterland).

Florida (2005) identifies the creative class as a driving force of contemporary cities and regions. The economic development of cities is dependent on the 3Ts, i.e. technology (innovation, high concentration of high-tech companies), talent (share of highly educated people, but also of the creative folks, recognized by the "Bohemian Index") and tolerance (openness to all ethnic groups, races or forms of life, captured for example through the "Gay Index") (Florida et al., 2011).

Following Jan Jerde, "community-oriented ideas for architecture" (Revis, 2008) help to create the city as an event in itself. So-called "architainment" (Klein, 2004) with the example of Las Vegas and landmark buildings (after the success of Frank O. Gehry's Bilbao Guggenheim called "the Bilbao effect"; Rybczynski, 2002) increase the attractiveness of the city.

Data from cell phones and sensor-based devices (e.g., CCTV cameras) and their rapid application in urban systems (for example, in traffic control) lead—according to Ratti, Frenchman, Pulselli, and Williams (2006)—to "mobile landscapes" and the "real-time city".

Libraries in cities of the knowledge society act as physical libraries (providing spaces for reading, learning or meeting and offering printed books as well as journals) and as digital libraries (offering all kinds of digital information) (Mainka, Hartmann, Stock, & Peters, 2013). Governments in such cities tend to communicate with their stakeholders via mature and usable websites (Fietkiewicz, Mainka, & Stock, 2017) as well as via social media (Mainka, Hartmann, Meschede, & Stock, 2015). Most of them publish their collected non-personal data ("open data") in order to arrive at more transparency and to provide for city-specific information services, e.g. mobile apps (Mainka et al., 2015a, b).

New and still nearly unstudied areas are the location factors in cities of the knowledge society. It is known from cities of the industry and the service society, that high per capita income and attractive living and working conditions are essential hard location factors for people. However, additionally there are soft location factors, for instance an attractive cityscape, many recreational and shopping facilities, schools and universities, libraries, etc. Location factors have importance for both, companies as well as people, to move into a city and to stay there.

For Vlacheas et al. (2013), the ultimate technological framework of smart cities is the Internet of Things (IoT), because it connects all components of the digital and some of the physical world. IoT technologies such as RFID (radio-frequency identification), embedded sensors and actuator nodes (Gubbi, Buyya, Marusic, &

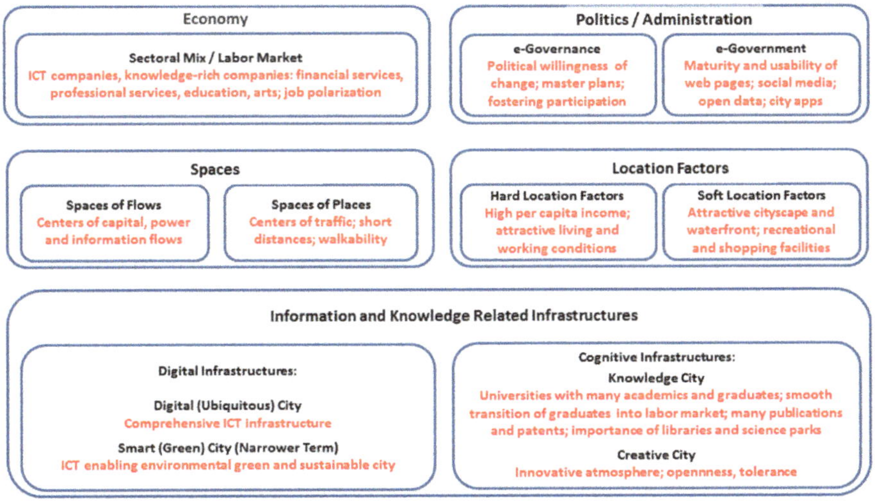

Fig. 1.3 Research framework

Palaniswami, 2013) are essential components of urban regions, making informational cities "ubiquitous".

An "informational city" is a typical city of the knowledge society. Here, all characteristics of the knowledge society appear on the city-level. Such a city is characterized by the two general aspects of cityness and informativeness (Fietkiewicz & Stock, 2014, 2015). Cityness includes building blocks which mainly exhibit the city's position within the network of all cities on the world, especially the indicators of the spaces of flows and the spaces of places as well as the location factors. Informativeness is directed on indicators of the information and knowledge related infrastructures, the knowledge-based sectoral mix and its labor market, the political willingness to create an informational city, to foster citizen participation and transparency as well as the maturity of e-government. All these elements discussed in smart city literature are summarized in Fig. 1.3, building the conceptual model for our research project. How can the status of a smart city in terms of its cityness and informativeness be measured?

1.3 Methods

In this chapter, we will report about our 34 interviews in the seven Gulf cities (Gremm, Barth, & Stock, 2015). We applied a quantitative questionnaire and conducted semi-standardized interviews with inhabitants between the 18th of February and the 6th of March 2013 in Abu Dhabi, Doha, Dubai and Manama and between the 28th of March and the 9th of April 2014 in Kuwait City, Muscat and

Sharjah. The duration of the interviews varied between 30 and more than 120 min. We based our questionnaire on the method of ServQual (Parasuraman, Zeithaml, & Berry, 1988) which is the standardized instrument to measure the quality of service in retailing and service organizations, but can be used in different research areas as well. We provided our interviewees with 18 questions that were split into two parts. The left hand side contains questions concerning the expectation (E) of the interviewees about an informational city in general and the right hand side contains the same question but refers to the experience (P) of the interviewees in the city in question. For each question there is a seven-ary scale where 7 represents the strongest form of agreement and 1 represents the strongest form of disagreement. To get an idea about the gap (Q) between the expectation and experience of a given question, the difference value $Q = E - P$ is calculated. For each question and for each city and also for the whole region, we calculate the arithmetic mean to create a ranking later on. We are aware of the fact that naturally Likert-type scales are ordinal and therefore do not allow using the arithmetic mean. However, since Likert (1932) reasons that consecutive natural numbers are equidistant, we interpret our scale as an interval scale and point this out to all our survey participants. Consequently we can calculate the arithmetic mean.

Since the original ServQual questions do not fit our needs, we designed our questionnaire with the help of literature. The question *Is the ICT infrastructure in an Informational City more important than automotive traffic infrastructure? (Q1)*, which belongs to the dimension of digital infrastructure, arose from the theory of the fifth Kondratieff according to which the economic wave of automobiles and petrochemistry is replaced by the wave of information technology (Nefiodow, 1991). The dimension of knowledge infrastructure contains three questions *Does an Informational City have to have knowledge-intensive companies and knowledge institutions, e.g. universities and science parks? (Q2)* This question was generated by thinking about Sassen's (2001) proposition that the city, where the headquarters of a company that operates on a global scale is located, becomes a center of information. For Mainka et al. (2013), besides the physical library space the offering of digital libraries is important, inducing the questions *Is a fully developed content infrastructure, e.g. supported by digital libraries, a characteristic feature of an Informational City? (Q3)* and *Are libraries important in an Informational City as a physical place for face-to-face communication? (Q4)*. For the creative infrastructure dimension we thought of the three T's mentioned by Florida (2002, 2005, 2012), Florida et al., (2011) and formulated the question *Does an Informational City have to be a creative city? (Q5)*. The T's refer to the terms technology, talent and tolerance, which are all required components for a city to be creative and economically successful (Murugadas, Vieten, Nikolic, Fietkiewicz, & Stock, 2015). We also included the questions *Are physical places where knowledge workers or creatives can meet each other important for an Informational City? (Q6)* in this dimension, since an informational city needs creative people who "want to meet and socialize with people unlike themselves, trade views and spare over issues" (Florida, 2002, p. 21). *Does an Informational City have to have an urban structure which is characterized by short distances? (Q7)* is an interesting question for the dimension of smart infrastructure, since the reduction of traffic causes better quality of life which allures creative people. "The key is to concentrate residences, work areas and amenities so as to produce the shortest possible trip distances, most being possible by bicycle and public transport" (Hall, 1997b, p. 15). Two questions belong to the dimensions of sectoral mix and labor market which we include together in one indicator. The first one is: *Does an Informational City have to have companies with information market activities, e.g. telecommunications companies? (Q8)*. Especially companies with information market activities should have their headquarters in an informational city. Furthermore, the question *Is the labor market in an Informational City characterized by job polarization (loss of routine middle-class jobs)? (Q9)* arose when we thought about the automatization of large economic processes and thus the occurrence of the two opposites (1) well-paid analytic tasks and (2) badly-paid manual labor (Goos & Manning, 2007). The dimension of political willingness contains two questions as well. Since the political willingness is an essential aspect to realize the idea of an informational city and to get financial support for the initial period, we asked: *Does the political willingness to create a knowledge economy play a significant role for establishing an Informational City? (Q10)*. The second question *Is an Informational City characterized by e-governance (incl. e-government, e-participation, e-democracy)? (Q11)* combines Moon's (2002) five steps of e-Government: (1) information dissemination (one-way communication), (2) two-way communication where the user gets a response, (3) service and financial transactions, (4) vertical and horizontal integration, and (5) political participation. The three questions *Does an Informational City have to have many leisure time facilities? (Q12)*, *Does an Informational City have to have an offer of a variety of shopping malls? (Q13)* and *Does an Informational City have to have architainment, e.g. spectacular skyline, amazing waterfront, big wheels (is such a city an event in itself)? (Q14)* deal with soft location factors which comprise the diversity of cultural and leisure activities. Hence, an informational city offers a variety of services, which cover different interests of individuals (Van den Berg, van der Meer, & Otgaar, 2007). Further three questions are generated to cover the dimension of spaces. The question *Does an Informational City have to have capital-intensive companies, e.g. banks and insurances? (Q15)* refers to Castells' (1989) space of flows. While formulating the question *Is an Informational City supposed to be a global city (world city)? (Q16)* we thought of "The World City Hypothesis" by Friedmann (1986, 1995) where cities are defined economically rather than by administrative borders. Given the fact that free access to information and freedom of communication are basic rights of anyone (Balkin, 2005), we asked: *Is it important for an Informational City that there is a free flow of all kinds of information (incl. mass media information)? (Q17)*. The last question *Do you think that your city is an Informational city? (Q18)* shall give an impression of our interview partner to which extend the discussed city corresponds to the prototypical informational city (Table 1.1).

Table 1.1 Informational cities in the GCC countries

Question/city	Kuwait city			Manama			Doha			Abu Dhabi		
	n = 5			n = 5			n = 4			n = 5		
	E	P	Q	E	P	Q	E	P	Q	E	P	Q
Q1: ICT versus automobile infrastructure	4.80	3.40	−1.40	5.20	4.00	−1.20	4.75	4.00	−0.75	5.20	4.40	−0.80
Q2: Knowledge-intensive facilities?	6.00	3.80	−2.20	6.60	4.80	−1.80	7.00	5.00	−2.00	6.80	5.80	−1.00
Q3: Digital libraries?	5.40	2.20	−3.20	6.80	3.20	−3.60	6.75	4.50	−2.25	6.20	1.20	−5.00
Q4: Physical libraries?	2.60	1.80	−0.80	5.40	3.60	−1.80	5.00	4.50	−0.50	5.00	2.20	−2.80
Q5: Creative city?	5.60	1.80	−3.80	6.25[3]	4.50	−1.75	6.50	5.00	−1.50	6.40	3.20	−3.20
Q6: Meeting places?	4.60	4.20	−0.40	6.80	5.60	−1.20	5.75	5.00	−0.75	5.80	3.00	−2.80
Q7: Short distances?	3.60	4.00	0.40	5.00	6.00	1.00	4.33[2]	4.33[2]	0.00	5.20	3.40	−1.80
Q8: Information market activities?	5.60	5.40	−0.20	6.40	6.00	−0.40	6.75	5.50	−1.25	6.40	4.20	−2.20
Q9: Job polarization?	4.20	3.00	−1.20	3.20	3.20	0.00	5.00[1]	5.67[2]	0.67	4.40	5.30	0.90
Q10: Political willingness?	6.00	4.80	−1.20	6.40	4.20	−2.20	6.25	6.25	0.00	5.80	3.75[3]	−2.05
Q11: EGovernance?	6.00	2.80	−3.20	5.60	5.40	−0.20	6.75	5.00	−1.75	6.20	4.00[3]	−2.20
Q12: Leisure facilities?	5.20	3.40	−1.80	5.40	4.40	−1.00	6.25	5.00	−1.25	5.80	4.80	−1.00
Q13: Shopping malls?	4.40	5.60	1.20	5.20	6.80	1.60	4.25	5.75	1.50	5.60	6.60	1.00
Q14: Spectacular buildings?	4.00	2.20	−1.80	5.20	4.00	−1.20	4.50	6.00	1.50	5.40	6.00	0.60
Q15: Capital-intensive companies?	5.20	5.60	0.40	6.20	6.00	−0.20	5.00	5.75	0.75	5.80	6.40	0.60
Q16: Global city?	3.80	2.10	−1.70	4.40	3.40	−1.00	6.00	5.75	−0.25	6.20	4.30	−1.90
Q17: Free flow of information?	6.60	4.20	−2.40	6.80	2.80	−4.00	6.75	4.50	−2.25	6.60	3.30	−3.30
Average for questions 1 to 17	**4.92**	**3.55**	**−1.37**	**5.70**	**4.58**	**−1.11**	**5.74**	**5.15**	**−0.59**	**5.81**	**4.23**	**−1.58**
Q18: Informational city?	–	**2.00**	–	–	**3.60**	–	–	**4.25**	–	–	**4.00**	–

(continued)

Table 1.1 (continued)

Question/city	Dubai			Sharjah			Muscat			GULF cities		
	n = 4			n = 5			n = 6			**n = 34**		
	E	P	Q	E	P	Q	E	P	Q	E	P	Q
Question/city	Dubai			Sharjah			Muscat			GULF cities		
	n = 4			n = 5			n = 6			**n = 34**		
	E	P	Q	E	P	Q	E	P	Q	E	P	Q
Q1: ICT versus automobile infrastructure	5.75	4.67[2]	−1.08	5.80	4.80	−1.00	5.33	2.00	−3.33	**5.26**	**3.79**	**−1.47**
Q2: Knowledge-intensive facilities?	6.75	5.83[1]	−0.92	6.40	5.80	−0.60	6.33	4.00	−2.33	**6.53**	**4.90**	**−1.63**
Q3: Digital libraries?	5.75	2.88	−2.88	6.00	4.40	−1.60	5.67	2.25	−3.42	**6.06**	**2.88**	**−3.18**
Q4: Physical libraries?	4.63	2.75	−1.88	6.00	5.40	−0.60	6.33	3.17	−3.16	**5.04**	**3.32**	**−1.72**
Q5: Creative city?	6.00	4.75	−1.25	6.20	4.80	−1.40	6.00	3.00	−3.00	**6.12**	**3.78**	**−2.34**
Q6: Meeting places?	6.50	5.25	−1.25	5.40	4.60	−0.80	6.67	3.50	−3.17	**5.94**	**4.38**	**−1.56**
Q7: Short distances?	5.25	4.00	−1.25	5.00	5.20	0.20	5.33	2.17	−3.16	**4.84**	**4.09**	**−0.75**
Q8: Information market activities?	6.25	4.50	−1.75	5.40	3.40	−2.00	4.67	5.50	0.83	**5.85**	**4.94**	**−0.91**
Q9: Job polarization?	4.00	4.25	0.25	4.20	3.60	−0.60	4.17	5.40[4]	1.23	**4.09**	**4.27**	**0.18**
Q10: Political willingness?	6.25	5.00[2]	−1.25	6.20	6.20	0.00	6.00	4.00	−2.00	**6.12**	**4.84**	**−1.28**
Q11: EGovernance?	6.00	4.00	−2.00	6.00	4.20	−1.80	5.33	3.20[4]	−2.13	**5.94**	**4.06**	**−1.88**
Q12: Leisure facilities?	6.50	5.75	−0.75	6.00	4.80	−1.20	6.17	4.17	−2.00	**5.88**	**4.56**	**−1.32**
Q13: Shopping malls?	6.25	6.50	0.25	5.20	4.60	−0.60	4.33	6.17	1.84	**5.00**	**6.00**	**1.00**
Q14: Spectacular buildings?	5.50	6.50	1.00	5.40	5.60	0.20	4.67	2.83	−1.84	**4.94**	**4.59**	**−0.35**
Q15: Capital-intensive companies?	6.25	5.00[2]	−1.25	5.20	4.80	−0.40	5.67	4.50	−1.17	**5.62**	**5.42**	**−0.20**
Q16: Global city?	6.25	5.50	−0.75	5.00	3.60	−1.40	3.50	1.83	−1.67	**4.91**	**3.62**	**−1.29**
Q17: Free flow of information?	6.50	4.00	−2.50	5.80	4.20	−1.60	6.67	1.83	−4.84	**6.53**	**3.46**	**−3.07**
Average for questions 1 to 17	**5.90**	**4.77**	**−1.13**	**5.60**	**4.71**	**−0.89**	**5.46**	**3.50**	**−1.96**	*5.57*	*4.29*	*−1.28*
Q18: Informational city?	–	**4.88**	–	–	**4.20**	–	–	**2.67**	–	–	*3.57*	–

P = Experience E = Expectation Q = P − E
n = Number of interviewees [1]: n = 2 [2]: n = 3 [3]: n = 4 [4]: n = 5
Results of the ServQual questionnaires

1.4 Informational Cities in the GCC Countries?

1.4.1 Digital City

Renowned as a part of the informational city, digital cities are based on ICT, which build a new kind of infrastructure (Fig. 1.4). Actually, there is no other technology that shapes social, cultural and economic change as much as ICT. The speed of these changes increases rapidly and transforms the world into an informational village where information and knowledge are provided and accessible for everyone. Place and time become irrelevant in a world where information can be transferred affordably and in a short time (Castells, 1989). In the economic sector, well-developed networks play a decisive role: Industry, energy, service and health sectors, all rely on it and would come to a standstill if they had to work without it. ICT permits them to act and move globally.

The average experience of the importance of ICT and its infrastructure was evaluated with averagely 3.79 points, which is 1.47 points below the expectation. ICT is broached in politics, but is almost unnoticeably implemented (interview partner DO1). The politicians mostly concentrate on road traffic due to the relevance of cars for the inhabitants, and the resulting traffic problems around the city (interview partners DO1, KU5). A life without cars is, particularly in Muscat

Fig. 1.4 Antenna hidden in an artificial palm tree, seen in Sharjah

(experience value: 2.00), almost inconceivable for the natives. Local public transport is—if provided—only used by foreign workers (interview partner AD3). The governments still invest in the expansion of road networks. In the private domain ICT seems to take a central position since otherwise the ways to communicate are limited. Some of the cities, namely Doha and those in the UAE seem to focus more on ICT than the others. On the whole, the internet connections are actually still too slow (interview partners DU3, KU2, MA3) and free WiFi hotspot availability is described as a rarity (interview partner MU3).

1.4.2 Knowledge City

While knowledge has always contributed to economic success, its importance has strikingly increased with the emergence of the information society and the knowledge society. Nowadays "cities are taking a leading role as both knowledge-consumption and knowledge-production hubs" (Carrillo et al., 2014, p. xvii). The new awareness of knowledge in all sectors paves the way for the knowledge city (Carrillo, 2006). One basic aspect of knowledge cities conveys "the conglomeration of intense scientific, technological, academic, cultural and innovation activities in urban spaces operating as engines of economic productivity" (Carrillo et al., 2014, p. 3). A second basic aspect of such cities includes well-educated and creative knowledge workers (Florida, 2005) and "creative communities" (Carrillo et al., 2014, p. 4) in order to establish knowledge-based urban development. Furthermore, the number of a city's libraries is an indicator "to assess the initial status of the city" (Ergazakis, Metaxiotis, & Psarras, 2006, p. 75) as a knowledge city, since they provide access to information. The term "knowledge cities" includes more than the pure existence of knowledge institutions and of knowledge workers in the city. All of "these elements constitute the cities' core identity, and the way their citizens use knowledge to build their social infrastructure, their institutions and their future" (Carrillo et al., 2014, p. xvii).

People living in the Gulf region see the importance of knowledge-intensive companies and knowledge institutions in an informational city and give an average of 6.53 points. The average experience is with 4.90 still below that score. The region is already working on the establishment of those institutions, but has still a long way to go (interview partner SH5). Knowledge institutions are available (Fig. 1.5), but knowledge-intensive companies are hardly present in Kuwait (interview partner KU4). Interviewees KU1 and MA3 voice misgivings of the knowledge-intensity of knowledge institutions. Even if they exist, they are not developed enough to be referred to as knowledge-intensive institutions. Many of the interviewees are of the opinion that the institutions and companies are not yet capable of building their own knowledge base. Neither in the institutions (interview partner MU1) nor in the companies (interview partner MU2) there is a sufficient research and development (Kosior, Barth, Gremm, Mainka, & Stock, 2015). For some national people living in the Gulf region, the reforms of the educational

Fig. 1.5 Knowledge Village in Dubai

systems are a "shock of the new" (Lightfood, 2014, p. 89), because such "modernist projects" confront the "traditional values in the Gulf" (Lightfood, 2014, p. 83).

A challenge for all Gulf countries is the human capital, which is able to master knowledge production as well as knowledge consumption in a knowledge society (Wiseman, Alromi, & Alshumrani, 2014). In the 1950s, in most of the Gulf States the foundations of a modern school system were laid. The first universities came decades later. Kuwait University was founded in 1962; the Gulf Technical College in Bahrain followed in 1968. In the Emirates, the UAE University in Al Ain was founded in 1976, and Zayed University, based in Abu Dhabi as well as in Dubai, began in 1998. The University of Qatar in Doha was established in 1973, and Muscat's Sultan Qaboos University opened in 1986. In the GCC countries, we find three kinds of higher education institutions: (1) public institutions (as, in the UAE, the UAE University, the Higher College of Technology, and the Zayed University); (2) local private institutions (as, e.g., the American University in Sharjah, the University of Dubai, and the Paris-Sorbonne University Abu Dhabi); and (3) international branch campuses (as, e.g. the campus of the University of Woolongong, Australia, which was the first international university, established in 1993).

A branch campus is "an offshore entity of a higher education institution operated by the institution or through a joint venture in which the institution is partner" (Becker, 2010, p. 3). In 2010, Becker reported on 162 branch campuses all over the world, which are dominated by institutions from the United States. Among the host countries, the UAE take the leading position. In the UAE, there are 40 international branch campuses; about two-thirds of these are located in Dubai Academic City (Becker, 2010, p. 4), the others in Abu Dhabi and in Sharjah. In Doha, we find 9 branch campuses, all of them located in the city district "Education City". Similar to other "cities" in the Gulf cities, in "Academic Cities", "Education Cities", "Knowledge Cities", etc. the branch campuses are 100% foreign ownerships; there is a 100% repatriation of profits and no tax (Becker, 2010, p. 4). In Doha's Education City, the Qatar Foundation bears all the costs of the infrastructures and of administrative assistance to the international universities.

The density of offshore campuses in the Gulf Region seems to be a regional phenomenon (Miller-Idriss & Hanauer, 2011), induced by the speed of transformation of the higher education systems in this region. The branch campuses lead towards "the Westernization—indeed the Americanization" of the Gulf cities' tertiary systems (Witte, 2010). On the other hand, we can identify some benefits of branch campuses. Since there are many international higher education institutions in the Gulf cities, more nationals, especially females, may decide to study at home instead of going abroad. It is possible to study, for example, at the New York University Abu Dhabi and to get a diploma from NYU. Because the students do not leave their country, the brain drain of Gulf nationals is reduced (Wilkins, 2011, p. 77). Although it would be possible for expatriates to attend a public university (with its specific rules, e.g. gender separation), most of the children of foreign families tend to study at international branch campuses (Wilkins, 2011, p. 78). In international universities, the instructional language is English; in public universities, the language depends on the subject taught. Courses in business administration, medicine, or engineering are mainly taught in English, while education, Shari'a (the Islamic legal system), and law are taught in Arabic (Findlow, 2006, p. 32).

University infrastructure at the Gulf has grown impressively during the last two decades. "Throughout the 1990s, in the Arab world, 10 new universities were established annually" (Donn & Al Manthri, 2010, p. 97). Thenceforth universities have to fulfill several roles: "ending illiteracy, diversification of hydrocarbon-based economies and preservation of cultural integrity" (Weber, 2011, p. 1). For instance, in Doha most of the universities grew up when Qatar Foundation's Education City was founded (Weber, 2014). This is an international campus, with about 1000 hectare located near the city that draws students from all over the world. There on the outskirts mostly international world-class universities are clustered on a large campus. Major American and European university branch campuses are partnered and established in this area.

Also in Sharjah, most of the universities and higher education institutions are positioned next to each other in an area called University City, which is a cluster for knowledge institutions (Interviewee MU3). Dubai builds up such areas likewise, including Dubai International Academic City and Dubai Knowledge Village. In total about 60 percent of Dubai's institutions are located inside one of those free trade zones.

As it turned out during the field research, in contrast to private universities some public universities in the Gulf Region even today teach gender-segregated, including Qatar University. Female and male bachelor students study separately and even have their own libraries. Collections are doubly present, but are mostly called up digitally (Interview partner DO1 and DO2). Indeed, nowhere else in the Gulf it is as strict as in Zayed University, one of the federal universities in Abu Dhabi: Female and male bachelor students are separated by a wall in the middle of the campus. This is due to local social conventions. Thus female students are not allowed to leave the campus without the approval of their parents, or if they are married, of their husband.

A huge problem of building up these knowledge areas is the great distance to the Gulf cities' centers, up to 30 km. Missing metro systems demand flexible mobility by the students and staff. The increasing number of students and the fact that the only way to arrive is by car causes traffic jams at Arab universities like Sultan Qaboos University (SQU) in Muscat (Abdul-Wahab & Fadlallah, 2014). Also, the German University of Technology in Oman (GUtech) is surrounded only by desert for few kilometers.

By funding of the Qatar Foundation the Qatar Science and Technology Park (QSTP) opened in 2009 inside the area of Education City. The research has a focus on energy, environment, health, and ICT and should support Qatar's post-oil and gas economy. QSTP is the only science park in the Middle East that is cooperatively located in a collective area with internationally prestigious universities. In 2012, about 50 companies (for example GE, Microsoft, and ExxonMobil) were based on these grounds. The out-of-town arcology project Masdar City, located next to Abu Dhabi, will prospectively include a Science and Technology Park. In addition, a technology park named CERT Technology Park that was founded by the Higher College of Technology, which is one of the public universities in Abu Dhabi, is located in the city. The exchange of technical knowledge and expertise should be promoted by this institution. Dubai houses the most range of science parks. The bioscience cluster Dubai Biotechnology & Research Park, the technology-oriented Techno Park, and the energy and environment park Enpark are indicators of an active development of science parks in Dubai. Additionally the technology park Internet City, harboring firms like Microsoft and IBM, and the business community Media City are located in Dubai's free trade zone next to Dubai Knowledge Village. Together they build up the first knowledge cluster in the region. Also in Muscat one science park is located: Knowledge Oasis Muscat (KOM). The park was established in 2005 and was built to encourage technology-oriented and thus knowledge-intensive businesses due to their increasing importance. The science park consists of over sixty companies, and even two IT-Colleges are located in the KOM.

There is a huge gap of 3.18 points between the average expectation (6.06) and the average experience (2.88) in case of the existence of digital libraries. Digital libraries include e-resources as well as digital reference services. Its content should be freely accessible for all inhabitants. Most of the interviewees are sure about the nonexistence of digital libraries. Some of them however include reference services without the corresponding full texts, which are offered to students by the universities (like Scopus) in the term ‘digital library’ (interview partners DU1, KU4, SH3). The use of the term therefore differentiates from the intended one.

For our interviewees the existence of physical libraries (5.04) is not as important as that of digital libraries. From the average experience (3.32), it can be deducted that not many libraries are present in the Gulf region. However, for the interviewees the problem lies elsewhere. The culture in the Gulf region is not a knowledge-based one. People do not get encouraged to read as they do not need to acquire knowledge for their jobs. Nearly all of the respondents agree that citizens do not use the available libraries (interview partners AD1, KU5, MA3). People only visit libraries for entertainment (interview partner KU4) or if they are somehow concerned with education, like teachers and students (interview partner SH5). However, the field research revealed that impressive libraries with great potential to attract visitors are available in the cities (Fig. 1.6). In Muscat, where the number of libraries is especially low, there are other possibilities to exchange books with other citizens (interview partner MU2).

One can find national libraries in all states except Oman, where such library is being planned. In Bahrain, the national library is part of the Isa cultural center. The university libraries meet the same standards as Western academic libraries. They not only present books on shelves, but ebooks and further digital media as well. We found impressive academic library buildings, e.g. in Qatar University or in Zayed

Fig. 1.6 Al-Babtain central library for Arabic poetry in Kuwait

University Abu Dhabi. Public libraries are not able to keep pace with institutions in other countries, say in Canada, China or Singapore (Mainka et al., 2013); in addition, they are poorly attended.

Librarians at the Gulf attach great importance to incorporate information literacy courses into their libraries. As an example, the American University of Sharjah Library "paid special attention to the design and placement of classroom facilities to ensure that the library's information literacy and instruction program, a critical component of the library's mission, could be delivered more effectively" (Ritchie & Ray, 2008, p. 167). All librarians are trained in information literacy skills. They are able to assist library users hereafter and to teach them MS Office Tools, library and IT basics like student folders, and library resources.

1.4.3 Creative City

In the twenty-first century a transformation of cities is happening. Cities have to find new ways to stabilize their economies for ensuring their economic power. Focusing on creativity seems like a good solution for today's societies (Chatterton, 2002). Also, researchers increasingly consider creativity as a huge benefit for cities, raising the concept of the creative city (Landry & Bianchini, 1995). According to Landry (2000, p. 4) the concept "describes a new method of strategic urban planning and examines how people can think, plan and act creatively in the city". He thinks that a creative city should offer cultural resources in order to generate and encourage creativity. Such resources include galleries, museums and places of cultural heritage (Fig. 1.7). Heritage areas and old city centers are able to echo the "spirit" of Arabian culture. Examples for such urban reconfiguration and revitalization (Wiedmann, Mirincheva, & Salama, 2013) are the Souk Waqif and the Msheireb project in Doha (Boussaa, 2014; Scharfenort, 2014). It is an open question, whether brand-new futuristic cityscapes as in Abu Dhabi, Doha, Dubai or Sharjah may contribute to the identity of the specific city as well (Boussaa, 2016).

Nowadays industries focus on knowledge production driven by creativity (Landry & Bianchini, 1995). Thus the existence of creative professionals who work in these knowledge-intensive industries is obligatory. A city should aim at attracting this "creative class" (Florida, 2002) by being tolerant and by offering enough employment opportunities and a high quality of living. In return, they are rewarded with more creative economic outcome, which is beneficial for their economic power.

With an average of 6.12, the interviewees think that in order to become an informational city, the city has to be creative. However, they do not feel positive about the creativity in the Gulf region, resulting in an average score of 3.78. The average value that is given by the Kuwaiti interviewees is with 1.80 even much lower than that. Interviewee KU3 is of the opinion that there is not enough

Fig. 1.7 Heritage village in Abu Dhabi

technology in Kuwait. Even though people try to apply it, they still have many problems with it. Since the talent of the citizens in the Gulf region is still undeveloped, the few talents that are available come from abroad (interview partners DU3, MU3). The region takes countermeasures to this problem by sending citizens to study abroad to expand their knowledge and abilities (interview partner MU3). A high tolerance threshold does not exist at the Gulf. There are many taboo topics like politics and homosexuality and partly also religion. According to interviewee MU1, Muscat is more tolerant than the rest of the region.

The respondents in the Gulf region think with an average score of 5.49 that knowledge workers and creative people must be offered places to communicate. The availability of such places in the Gulf region is rated with an average of 4.38. Especially in Abu Dhabi there seems to be a lack of meeting places. Here the average experience is only 3.00. The interviewees in Kuwait mentioned that the so called Dewaniya is a traditional meeting place like a tent or a house where people get together to cultivate contacts. Also malls, cafes, and restaurants are favored places to meet, but it is unknown if they are used to share and create knowledge (interview partners DU3, KU4). Interviewee SH3 states that meeting places become less important in advanced cities, but that in Sharjah these physical spaces are important since the city is still in the learning process.

1.4.4 Sustainable Green City

The smart city in the narrower sense is a green city including ICT aiming at a sustainable economy and better quality of life. Many GCC states invest in projects of construction of new "smart" cities, e.g., Abu Dhabi in Masdar City or Qatar in Lusail City (Tok, Al Mohammad, & Al Merekhi, 2014). The planning of those cities considers the application of ICT in the city (e.g., smart transportation) as well as in the homes (e.g., cameras, lighting, sprinklers, and security systems), sustainability (zero carbon emissions in Masdar) and traditional Arabic architecture. For Masdar, Tok et al. (2014, p. 136) pinpoint that such a smart Arabian city "does have a remarkable role in (the) transition from carbon-based economy to sustainable economy. The climate change has also created a global demand for alternative sources of energy. Hence, Masdar City as the vital technological cluster in the region will make Abu Dhabi a global energy leader." As both cities are in the construction phase, it is an open question, whether they will be accepted by companies, residents, and tourists.

With an average of 4.34, the interviewees in the Gulf cities think that short distances should be realized in an informational city. Whether the citizens of the Gulf region can reach their destinations by short routes is estimated with an average score of 4.09. Especially the opinion of the interview partners in Manama stands out. They think with an average of 6.0 that Manama is characterized by short distances. In most of the cities at the Gulf, the offer of public transport is scarce. Due to our field research, it can be said that the investment in public transport has no visible effect. Except for the metro in Dubai, which is accordingly a widely used mean of transportation (interview partner DU2), no railway or metro system exists. People resort to using cars which results in high traffic density and people being delayed (interview partner KU2). According to our interview partners in Muscat, a short urban structure does not exist in the capital area. Muscat expanded its urban area in recent years and is nowadays an urban conglomeration stretching 60 km rendering it impossible for Muscat to offer short distances. The long ways from one point of the capital area to another is also what we witnessed during our field research. In the past, Muscat had clusters and consequently short distances, but nowadays the companies and population are spread throughout the whole city (interview partner MU2). Mattrah (one of the old city clusters of Muscat) is still characterized by short distances in terms of walkability and cyclability (Fig. 1.8).

1.4.5 Economy

In a knowledge society, knowledge-intensive industries, capital-intensive industries and ICT industries are corresponding factors on the labor market. In this case, the term 'knowledge-intensive industries' means science parks, consultants, etc., 'capital-intensive industries' refers to banks and insurance companies whereas ICT

Fig. 1.8 Rent a bike at Mattrah's corniche

industries are, for example, telecommunication companies and information services. Not least, creative industries, in terms of Florida's (2012) Bohemian Index, which measures the density of artists, writers and performers in a region, are a component of a city's labor market in the knowledge society. The workers, which are employed in the industries of the knowledge society, work rather with their brain, than with their hand.

The inhabitants of the Gulf cities answered the question concerning the necessity of companies with information market activities averagely with 5.85 points on the expectation and 4.94 on the experience side. The difference between the assigned scores of the single cities is conspicuous: Doha—having an own telecommunication company—gets 6.75 points for expectation and 5.50 points for experience and others like, for example, Sharjah—having no own telecommunication companies and depending on the neighboring regions Abu Dhabi and Dubai (interview partner SH4)—gets only 3.40 points for the experience part. In this context, also some powerful telecommunication companies like Etisalat located in the UAE, Zain located in Kuwait City and Nawras and Omantel located in Muscat were named (interview partners KU1, MU1). Often, companies are located in centered regions like "Internet City" and "Media City" in Dubai (interview partner AD3). Nevertheless, the telecommunication market is seen as rather small (interview partner DO1). The interviewees agree with each other that their services are insufficient. Interview partner MU2 mentions that the providers are just specialized in sale and, hence, are non-manufacturing companies. In contrast, a software branch does not exists is the Arab region at all (interview partners DO1, MA4).

The most striking feature concerning the knowledge workers is that there are mostly more non-citizens than citizens employed at knowledge-intensive institutions. Following KU4, the majority of the academic staff at private institutions in Kuwait City are foreigners and citizens would rather work at public universities. Natives do not work at private universities and especially not at the branch universities located in the Gulf Region. For many years in a row now, nearly all of the academics in private or branch universities of the cities in the Gulf Region have been foreigners. Most of the academics are expat Arabs who have a Ph.D. from the United States or Britain.

Averagely the interview partners answered the question regarding job polarization with 4.27 points on the experience side, which is 0.18 points higher than their expectation. Kuwaitis, Emirati and Omanis speak about a strict separation between high and low paid jobs, but also emphasize the importance of the existing middle class. Many of these jobs are indispensable (interview partners KU4, MU1, SH1). In contrast to most of the other natives, Omanis work beside the expats as common professions (interview partner MU1). The expat population in the Gulf cities—most of them live in the country for work—is almost everywhere above 80 percent except in Muscat, where it is a bit lower. Nationalization programs like 'Qatarization' or 'Emiratization' are in progress to reduce the high number of expats (interview partners DO1, KU1). It is doubtful whether this question is appropriate for the Gulf region since most of the natives work in the public middle class sector (Kosior et al., 2015).

1.4.6 Politics/Administration

Political willingness and e-Government are required conditions to establish an informational city. E-Government refers to the use of ICT and especially the inclusion of web-based services by the government. This form of government should be established primarily for the citizens since it provides a relief in governmental matters like the access to information, hold available by government (Awan, 2003).

The political willingness to perform changes exists in all of the analyzed cities (average experience rating: 4.84) and their future visions are supported by large sums of money. The willingness seems to be particularly high in Doha (6.25), Sharjah (6.20), and Dubai (5.00), however readiness to cooperate on a regional level only exist to a minor degree (interview partner DO1). The interviewees mention, that the global trend of city development is recognized and that plans to change the future of the Gulf states exist. The interviewees in Dubai confirm that the city "has the willingness of contesting the world" (interview partner DU1) and therefore tries to improve their status. Other Emirati take the prototypical informational city Singapore as an example for their future urban development towards a knowledge society (interview partner AD2). The future plans will be financed by the oil and gas revenues. However, in most cases the implementation proves difficult (interview partners KU3, MU5). Interviewee KU3 predicts that someday, when the oil reserves are empty, the government will be unemployed and the country will be empty since all people will leave.

When asked if Kuwait City, Manama, Doha, Abu Dhabi, Dubai, Sharjah and Muscat are characterized by e-Governance, the interviewees give averagely 4.06 points. Thus, their experience is 1.88 points lower than their expectation of e-Governance in an informational city. This difference is accounted for by the existence of e-Government and e-Participation, and by the lack of e-Democracy (interview partner SH2). Information is available, but the citizens do not get the chance to participate in governmental matters, since all decisions are already made (interview partner KU1). Only some transactions can be done online, but one thing seems to be possible in all Gulf cities: the payment of parking tickets online (interview partners AD1, DU3, MA1). Interviewee KU3 talks about improvable services, such as the mail service of the government, while in Muscat up to 90 percent of the services can already be done online (interview partner MU3). Even though activity and participation on social media platforms like Facebook and Twitter exist, it is not always sufficient (interview partners SH3, MU2).

1.4.7 Location Factors

In an informational city, central aspects are the location factors, especially the soft location factors. The presence of several of these factors contributes to the attractiveness and the increase of the magnetic effect of a city. Soft location factors comprise cultural and leisure time facilities like museums, galleries, libraries, opera houses, symphony halls, theatres, and diverse events, architainment and attractive waterfronts as well as shopping opportunities (Hall, 1997a). In order to attract inhabitants, business companies and visitors, a city needs to offer a variety of those services to cover the different interests of the individual groups (Van den Berg et al., 2007).

The presence of leisure time facilities in the Gulf region is rated 4.56 and is therefore not as high as the expectation (5.88). The average experience value of the interviewees in Dubai is quite high. But not only do the residents think highly about the available leisure time facilities in Dubai but also interview partner SH4 confessed that residents of Sharjah go to Dubai for the consumption of leisure facilities.

The question about shopping malls is one of the two questions where the average expectation (5.00) lies below the average experience (6.00). All the interviewees decided unanimously that there are enough or even more than enough shopping

Fig. 1.9 Architainment as a soft location factor: Burj Khalifa area in Dubai

malls in the Gulf region. These malls are not only used for the purpose of shopping but they serve as a social place for the people. Due to extremely high temperatures in summer, the climate in the malls is pleasant for the residents (interview partner MU4). Again, it holds for Sharjah that even though the emirate has enough shopping possibilities (interview partner SH3), people prefer to go to Dubai or Abu Dhabi instead (interview partner SH4).

The average expectation regarding the existence of architainment (Fig. 1.9) in the Gulf cities is 4.94, which is only 0.35 above the average experience. The average experiences in Muscat and especially in Kuwait are very low. Kuwait is accordingly not attractive regarding architainment and waterfronts. The cities' attractiveness is limited to light effects since there is no ancient substance (interview partner KU1). In Muscat the construction of buildings with more than eight floors is forbidden by Oman's building law (interview partner MU1). There are no other contemporary or sophisticated buildings either. Muscat can score with its very nice nature made up of mountains, beaches, wadis (dry riverbeds), and also desert landscapes; however there are only few tourists (interview partner MU2). Unlike Muscat, Dubai attracts lots of tourists every year. Some of them even come to visit more than once (interview partner DU2). In Doha and Abu Dhabi the average experience is just a bit lower than in Dubai.

1.4.8 Spaces

First efforts researching world cities were made by "The World City Hypothesis" (Friedmann, 1986), which deals with the city's placement in the world economy. Castells' (1989) city research is also included and defines cities by their placement in the space of flows: flows of money, power, and information compared to the former focus on space of places. According to Taylor (2004) world cities are never isolated but always interconnected. The Gulf cities seem to have a sufficient number of capital-intensive companies, since the expectation (5.62) and the experience (5.42) almost achieved the same valuation. Some of the cities have banking districts (interview partner MU2) (Fig. 1.10) where the British finance system prevails. Islamic banking—a system where no interests are charged—also co-exists, but since it is suspicious to many people, some cities, e.g. Muscat, refused to introduce this system for a long a time (interview partner MU3).

Regarding the aspect of informational cities and Gulf cities as global cities, the expectations were averagely higher (4.91) than the experiences (3.62). At it, interviewees in Doha (5.75), Dubai (5.50) and Abu Dhabi (4.30) are most confident of living in a global city. Especially in Dubai it was emphasized that the city is well-connected (interview partner DU2). Others, for instance living in Sharjah, mention that their city is trying to become global (interview partner SH4). And still

Fig. 1.10 Bahrain Financial Harbor in Manama

others, living in Kuwait, Manama or Muscat, do not classify their city as global. A global city has to be well connected and needs to attract people (interview partner KU1)—however, Muscat is called a “boring village” (MU1).

For Sigler (2013), “relational cities” intermediate between global and regional economies. They have containerized and bulk-specialized transportation infrastructures (e.g., oil or gas harbors), good shipping and air transport networks, apply fiber optics as well as satellite networks for communication and practice advanced capitalism. Sigler calls Dubai as well as Doha “relational cities”, as they are hubs in the world trade (especially between Europe and Asia) and in the trade between the world and the regional Arab economies.

In the Gulf states, a strict policy of censorship is pursued (average experience value: 3.46): Voice over IP (VoIP) is officially banned. However, the Skype website could be opened in all cities, but not in Muscat. The government tries to control every sector to operate their political power (interview partners DU2, MU2). Also, it can be assumed that the telecommunication companies do not want to compete with freely available software and that they want to generate more profit (KU4). However, a virtual private network (VPN) client is widely used to get access to Skype and other censored websites (interview partners MA3, KU4). The interviewees feel restricted in expressing their own opinion (interview partner MU4). The expected necessity of free flow of information in an informational city (average expectation value: 6.53) cannot be confirmed for the Gulf cities.

1.4.9 Ranking of the Gulf Cities

The final point of our research about the Gulf cities’ cityness and informativeness is an overall ranking of Kuwait City, Manama, Doha, Abu Dhabi, Dubai, Sharjah, and Muscat (Table 1.2). In case of the interviewees’ experiences (P), Doha receives the

Table 1.2 Ranking of GCC cities by the ServQual results

Ranking by P	P-value	Ranking by Q	Q-value	Ranking by Q18	Q18-value	Ranking by position	Position value
Doha	5.15	Doha	−0.59	Dubai	4.88	Doha	*4*
Dubai	4.77	Sharjah	−0.89	Doha	4.25	Dubai	*7*
Sharjah	4.71	Manama	−1.11	Sharjah	4.20	Sharjah	*8*
Manama	4.58	Dubai	−1.13	Abu Dhabi	4.00	Manama	***12***
Abu Dhabi	4.23	Kuwait city	−1.37	Manama	3.60	Abu Dhabi	***15***
Kuwait City	3.55	Abu Dhabi	−1.58	Muscat	2.67	Kuwait City	***18***
Muscat	3.50	Muscat	−1.96	Kuwait City	2.00	Muscat	***20***
Gulf City	**4.29**	**Gulf city**	**−1.28**	**Gulf City**	**3.57**		

best score with 0.86 points above the average of the evaluated Gulf cities and is followed by Dubai and Sharjah. Abu Dhabi, Kuwait City and Muscat fall below the average. Regarding the difference (Q) between experience and expectation, Doha—where the interviewees' experiences do not deviate much from their expectations—comes first once more. Also Sharjah achieves again a good rank. The values of Kuwait City, Abu Dhabi and Muscat are once again below the average. Taking the last question on informational cities (Q18) into consideration the huge difference between the best (Dubai: 4.88) and the worst (Kuwait: 2.00) rated city is striking. To receive the final ranking of all cities, the positions of the rankings by P, Q and question 18 are aggregated: Doha in Qatar is ranked first and hence has the highest cityness and informativeness regarding the estimations of our interview partners. It is followed by Dubai and then Sharjah. The fourth and fifth positions are occupied by Manama and Abu Dhabi while Kuwait and Muscat come in last.

For Kuwait City—a city concentrating rather on the development of automotive infrastructure than the development of ICT—an after-oil future with an empty cityscape is predicted. Gone are the times when Kuwait was attractive to foreigners due to its prosperity. Having impressive libraries that are barely used by the citizens, meeting places where people rather talk about other things than knowledge-intensive topics and no architainment to attract people from different countries seem to be determining disadvantages in the struggle to become a successful informational city. It gets of barely better than Muscat where receiving the status of an informational city seems far away.

In contrast, Dubai, as the most known of the seven United Arab Emirates, out-shining Abu Dhabi and Sharjah in terms of informativeness, focuses its strategy on ICT more than other cities in the region. The construction of an own metro system, the settlement of ICT companies and technology parks testify a huge development during the last few years. But will the city be superseded in the future? According to our ServQual results, the willingness to become an informational city seems to be highest in Doha. The development of an ICT infrastructure is as high as in Dubai and also some ICT companies are located in the city. The plan is to educate and train nationals properly to pave the way to a global knowledge society. This proves that the willingness to pursue the status of an informational city definitely exists. The will is also existent in Manama, which is ranked exactly in the middle of all cities, but the success has, unlike Doha, not occurred. The Arabian oryx is Qatar's national animal. "The Oryx takes off", Conventz, Thierstein, Salama, and Wiedmann (2015, p. 65) state. "The oryx antelope is the iconic logo of Qatar Airways, which can be regarded as one epitome of Doha's being a rising hub in the Gulf region". To sum up, we can carefully say that Kuwait is the past, Dubai is the present and Doha is the future (Gremm, Barth, & Stock, 2015). That is why we are going to study Doha and the state of Qatar in more detail.

References

Abdul-Wahab, S. A., & Fadlallah, S. O. (2014). A study of the effects of vehicle emissions on the atmosphere of Sultan Qaboos University in Oman. *Atmospheric Environment, 98,* 158–167.

Albino, V., Berardi, U., & Dangelico, R. M. (2015). Smart cities: Definitions, dimensions, performance, and initiatives. *Journal of Urban Technology, 22*(1), 3–21.

Awan, M. (2003). E-Government: Assessment of GCC (Gulf co-operating Council) countries and service provided. *Lecture Notes in Computer Science, 2739,* 500–503.

Balkin, J. M. (2005). Digital speech and democratic culture: A theory of freedom of expression for the information society. In A. D. Moore (Ed.), *Information ethics, privacy, property, and power* (pp. 297–354). Seattle, WA: University of Washington Press.

Barth, J., Fietkiewicz, K. J., Gremm, J., Hartmann, S., Ilhan, A., Mainka, A., et al. (2017). Informational urbanism. A conceptual framework of smart cities. In *Proceedings of the 50th Hawaii International Conference on System Sciences, January 4–7, 2017, Waikoloa Village* (pp. 2814–2823). Washington, DC: IEEE Computer Society.

Becker, R. (2010). International branch campuses: New trends and directions. *International Higher Education, 58,* 3–5.

Boussaa, D. (2014). Al Asmakh historic district in Doha: From an urban slum to living heritage. *Journal of Architectural Conversation, 20*(1), 2–15.

Boussaa, D. (2016). Cities in the Gulf: Rapid urban development and the search for identity in a global world. In U. G. Benna & S. B. Garba (Eds.), *Population growth and rapid urbanization in the developing world* (pp. 166–191). Hershey, PA: IGI Global.

Caragliu, A., Del Bo, C., & Nijkamp, P. (2011). Smart cities in Europe. *Journal of Urban Technology, 18*(2), 65–82.

Carrillo, F. J. (Ed.). (2006). *Knowledge cities: Approaches, experiences, and perspectives.* New York, NY: Butterworth Heinemann.

Carrillo, F. J., Yigitcanlar, T., García, B., & Lönnqvist, A. (2014). *Knowledge and the city. Concepts, applications and trends of knowledge-based urban development.* New York, NY: Routledge.

Castells, M. (1989). *The informational city. Information technology, economic restructuring, and the urban-regional process.* Oxford, UK, Cambridge, MA: Blackwell.

Castells, M. (1996). *The rise of the network society.* Malden, MA: Blackwell.

Castells, M. (1997). *The power of identity.* Malden, MA: Blackwell.

Castells, M. (1998). *End of millennium.* Malden, MA: Blackwell.

Chatterton, P. (2002). Will the real creative city please stand up? *City, 4*(3), 390–397.

Chourabi, H., Nam, T., Walker, S., Gil-Garcia, J. R., Mellouli, S., Nahon, K., et al. (2012). Understanding smart cities: An integrative framework. In *Proceedings of the 45th Hawaii International Conference on System Sciences* (pp. 2289–2297). Washington, DC: IEEE Computer Society.

Cocchia, A., (2014). Smart and digital city: A systematic literature review. In R. P. Dameri & C. Rosenthal-Sabroux (Eds.), *Smart city. How to create public and economic value with high technology in urban space* (pp. 13–43). Cham, Switzerland: Springer.

Conventz, S., Thierstein, A., Salama, A. M., & Wiedmann, F. (2015). When the Oryx takes off: Doha a new rising knowledge hub in the Gulf-region? *International Journal of Knowledge-Based Development, 6*(1), 65–82.

Cook, D. J. (2012). How smart is your home? *Science, 335*(6076), 1579–1581.

Donn, G., & Al Manthri, Y. (2010). *Globalisation and higher education in the Arab Gulf States.* Didcot, UK: Symposium.

EIU. (2009). *The GCC in 2020: The gulf and its people.* Dubai, AE: The Economist Intelligence Unit Limited.

EIU. (2010). *The GCC in 2020: Resources for the future.* Dubai, AE: The Economist Intelligence Unit Limited.

Ergazakis, K., Metaxiotis, K., & Psarras, J. (2006). An emerging pattern of successful knowledge cities' main features. In F. J. Carrillo (Ed.), *Knowledge cities: Approaches, experiences, and perspectives* (pp. 3–15). New York, NY: Butterworth Heinemann.

Fietkiewicz, K. J., Mainka, A., & Stock, W. G. (2017). Egovernment in cities of the knowledge society. An empirical investigation of smart cities' governmental websites. *Government Information Quarterly, 34*(1), 75–83.

Fietkiewicz, K. J., & Stock, W. G. (2014). Cityness and informativeness of the emerging informational cities in Japan. *Creative and Knowledge Society, 4*(1), 43–56.

Fietkiewicz, K. J., & Stock, W. G. (2015). How 'smart' are Japanese cities? An empirical investigation of infrastructures and governmental programs in Tokyo, Yokohama, Osaka and Kyoto. In *Proceedings of the 48th Hawaii International Conference on System Sciences* (pp. 2345–2354). Washington, DC: IEEE Computer Science.

Findlow, S. (2006). Higher education and linguistic dualism in the Arab Gulf. *British Journal of Sociology of Education, 27*(1), 19–36.

Florida, R. L. (2002). The rise of the creative class: Cities without gays and rock bands are losing the economic development race. *The Washington Monthly, 34*(5), 15–25.

Florida, R. L. (2005). *Cities and the creative class*. New York, NY, London, UK: Routledge.

Florida, R. L. (2012). *The rise of the creative class: revisited*. New York, NY: Basic Books.

Florida, R. L., et al. (2011). *Creativity and prosperity: The global creativity index*. Toronto, ON: Martin Prosperity Institute.

Foth, M., Choi, J. H.-H., & Satchell, C. (2011). Urban informatics. In *Proceedings of the ACM 2011 Conference on Computer Supported Work* (pp. 1–8). New York, NY: ACM.

Friedmann, J. (1986). The world city hypothesis. *Development and Change, 17,* 69–83.

Friedmann, J. (1995). Where we stand. A decade of world city research. In P. Knox & P. Taylor (Eds.), *World cities in a world-system* (pp. 21–47). Cambridge, UK, New York, NY: Cambridge Univ. Press.

GCCSG. (2012). *The cooperation council for the Arab States of the Gulf.* Retrieved from www.gcc-sg.org.

Giffinger, R., Fertner, C., Kramar, H., Kalasek, R., Pichler-Milanovic, N., & Meijers, E. (2007). *Smart cities—ranking of European medium-sized cities*. Vienna, Austria: Centre of Regional Science.

Goos, M., & Manning, A. (2007). Lousy and lovely jobs: The rising polarization of work in Britain. *Review of Economics and Statistics, 89*(1), 118–133.

Gremm, J., Barth, J., & Stock, W. G. (2015). Kuwait is the past, Dubai is the present, Doha is the future: Informational cities on the Arabian Gulf. *International Journal of Knowledge Society Research, 6*(2), 51–64.

Gubbi, J., Buyya, R., Marusic, S., & Palaniswami, M. (2013). Internet of Things (IoT): A vision, architectural elements, and future directions. *Future Generation Computer Systems, 29,* 1645–1660.

Hall, P. (1966). *The world cities*. London, UK: Weidenfeld and Nicolson.

Hall, P. (1997a). Modelling the post-industrial city. *Futures, 29*(4/5), 311–322.

Hall, P. (1997b). *The first megacity lecture: Megacities, world cities and global cities*. Rotterdam, NL: Stichting Megacities.

Hall, R. E., Bowerman, B., Braverman, J., Taylor, J., Todosow, H., & von Wimmersperg, U. (2000, September 28). The vision of a smart city. In *2nd International Life Extension Technology Workshop, Paris, France*.

Hollands, R. G. (2008). Will the realsmart city please stand up? *City, 12*(3), 303–320.

Jacobs, J. (1969). *The economy of cities*. Random House, New York, NY: Random House.

Klein, N. M. (2004). *The Vatican to Vegas. A history of special effects*. London, UK, New York, NY: New Press.

Kosior, A., Barth, J., Gremm, J., Mainka, A., & Stock, W. G. (2015). Imported expertise in world-class knowledge infrastructures: The problematic development of knowledge cities in the Gulf region. *Journal of Information Science Theory and Practice, 3*(3), 17–44.

Kourtit, K., Nijkamp, P., & de Noronha Vaz, T. (2015). Cities in a shrinking globe. *International Journal of Global Environmental Issues, 14*(1/2), 6–16.

Kumar, K. B., & van Welsum, D. (2013). *Knowledge-based economies and basing economies on knowledge: Skills a missing link in GCC countries*. Santa Monica, CA: RAND Corporation.

Landry, C. (2000). *The creative city: A toolkit for urban innovation*. London, UK: Earthscan.

Landry, C., & Bianchini, F. (1995). *The creative city*. London, UK: Demos.

Lightfood, M. (2014). Building a knowledge society on sand: When the modernist project confronts the traditional values in the Gulf. In A. W. Wiseman, N. H. Alromi, & S. Alshumrani (Eds.), *Education for a knowledge society in Arabian Gulf countries* (pp. 83–101). Bingley, UK: Emerald.

Likert, R. (1932). A technique for the measurement of attitudes. *Archives of Psychology, 22*(140), 3–55.

Madanipour, A. (2011). *Knowledge economy and the city. Spaces of knowledge*. London, UK, New York, NY: Routledge.

Mainka, A., Hartmann, S., Meschede, C., & Stock, W. G. (2015a). Mobile application services based upon open urban government data. In *Proceedings of iConference 2015. Newport Beach CA, USA, March 24–27, 2015* (15 pp.).

Mainka, A., Hartmann, S., Meschede, C., & Stock, W. G. (2015b). Open government: Transforming data into value-added city services. In M. Foth, M. Brynskov, & T. Ojala (Eds.), *Citizen's right to the digital city. Urban interfaces, activism, and placemaking* (pp. 199–214). Singapore, SG: Springer.

Mainka, A., Hartmann, S., Orszullok, L., Peters, I., Stallmann, A., & Stock, W. G. (2013). Public libraries in the knowledge society: Core services of libraries in informational world cities. *Libri, 63*(4), 295–319.

Mainka, A., Hartmann, S., Stock, W. G., & Peters, I. (2015b). Looking for friends and followers: A global investigation of governmental social media use. *Transforming Government: People, Process and Policy, 9*(2), 237–254.

Miller-Idriss, C., & Hanauer, E. (2011). Transnational higher education: Offshore campuses in the Middle East. *Comparative Education, 47*(2), 181–207.

Mitchell, W. J. (1995). *City of bits. Space, place, and the infobahn*. Cambridge, MA: MIT.

Mitchell, W. J. (1999). *e-topia. Urban life, jim—but not as we know it*. Cambridge, MA: MIT.

Mitchell, W. J. (2003). *Me++. The cyborg-self and the networked city*. Cambridge, MA: MIT.

Moon, M. J. (2002). The evolution of e-government among municipalities: Rhetoric or reality? *Public Administration Review, 62*(4), 424–433.

Murugadas, D., Vieten, S., Nikolic, J., Fietkiewicz, K. J., & Stock, W. G. (2015). Creativity and entrepreneurship in informational metropolitan regions. *Journal of Economic and Social Development, 2*(1), 14–24.

Nam, T., & Pardo, T. A. (2011a). Conceptualizing smart city with dimensions of technology, people, and institutions. In *Proceedings of the 12th Annual International Conference on Digital Government Research* (pp. 282–291). New York, NY: ACM.

Nam, T., & Pardo, T. A. (2011b). Smart city as urban innovation: Focusing on management, policy, and context. In E. Estevez & M. Janssen (Eds.), *5th international conference on theory and practice of electronic governance* (pp. 185–194). New York, NY: ACM.

Nefiodow, L. A. (1991). *Der fünfte Kondratieff: Strategien zum Strukturwandel in Wirtschaft und Gesellschaft*. Wiesbaden, Frankfurt/Main, Germany: Gabler, FAZ.

Negre, E., & Rosenthal-Sabroux, C. (2015). Smart cities: A salad bowl of citizens, ICT and environment. In A. Vesco & F. Ferrero (Eds.), *Handbook of research on social, economic, and environmental sustainability in the development of smart cities* (pp. 61–78). Hershey, PA: IGI Global.

Neirotti, P., De Marco, A., Cagliano, A. C., Mangano, G., & Scorrano, F. (2014). Current trends in smart city initiatives: Some stylised facts. *Cities, 38*, 25–36.

O'Grady, M., & O'Hare, G. (2012). How smart is your city? *Science, 335*(1579), 1581–1582.

Parasuraman, A., Zeithaml, V. A., & Berry, L. L. (1988). SERVQUAL: A multiple-item scale for measuring consumer perceptions of service quality. *Journal of Retailing, 64*(1), 12–40.

Ratti, C., Frenchman, D., Pulselli, R. M., & Williams, S. (2006). Mobile landscapes: Using location data from cell phones for urban analysis. *Environment and Planning B: Planning and Design, 33*(5), 727–748.

Revis, J. (2008, Spring/Summer). The art of architecture, Interview with Jon Jerde. *Art and Living*, 69–76.

Ritchie, L., & Ray, K. (2008). Incorporating information literacy into the building plan: The American University of Sharjah experience. *Reference Services Review, 36*(2), 167–179.

Rybczynski, W. (2002). The Bilbao effect. *The Atlantic, 290*(2), 138–142.

Salama, A. M., & Wiedmann, F. (2013). *Demystifying Doha*. Farnham, UK: Ashgate.

Sassen, S. (2001). *The global city. New York, London, Tokyo* (2nd ed.). Princeton, NJ: Princeton University Press.

Scharfenort, N. (2014). The Msheireb project in Doha: The heritage of new urban design in Qatar. In K. Exell & T. Rico (Eds.), *Cultural heritage in the Arabian Peninsula* (pp. 189–204). Milton Park, UK: Routledge.

Shin, D. H. (2009). Ubiquitous city. Urban technologies, urban infrastructure and urban informatics. *Journal of Information Science, 35*(5), 515–526.

Sigler, T. J. (2013). Relational cities: Doha, Panama City, and Dubai as 21st century entrepôts. *Urban Geography, 34*(5), 612–633.

Simmel, G. (1903). Die Großstädte und das Geistesleben. In *Die Großstadt. Jahrbuch der Gehe-Stiftung, Vol. 9* (pp. 185–206).

Søderström, O., Paasche, T., & Klauser, F. (2014). Smart cities as corporate storytelling. *City, 18* (3), 307–320.

Stallmeyer, J. C. (2009). Landscapes of informational urbanism. *Journal of Landscape Architecture, 6*(3), 34–39.

Stallmeyer, J. C. (2011). *Building Bangalore. Architecture and urban transformation in India's Silicon Valley*. London, UK, New York, NY: Routledge.

Stock, W. G. (2011). Informational cities. Analysis and construction of cities in the knowledge society. *Journal of the American Society for Information Science and Technology, 62*(5), 963–986.

Stock, W. G. (2015). Informational urbanism. *Journal of Systemics, Cybernetics and Informatics, 13*(6), 62–69.

Stock, W. G., & Stock, M. (2013). *Handbook of information science*. Berlin, Germany, Boston, MA: De Gruyter Saur.

Tadros, M. E. (2015). *The Arab Gulf States and the knowledge economy: Challenges and opportunities*. Washington, DC: Arab Gulf States Institute.

Taylor, P. J. (2004). *World city network. A global urban analysis*. London, UK: Routledge.

Taylor, P. J., Hoyler, M., & Verbruggen, R. (2010). External urban relational process. Introducing central flow theory to complement central place theory. *Urban Studies, 47*(13), 2803–2818.

Tok, E., Al Mohammad, F., & Al Merekhi, M. (2014). Crafting smart cities in the Gulf region: A comparison of Masdar and Lusail. *European Scientific Journal, 2*, 130–140.

Ulrichsen, K. C. (2016). *The Gulf States in international political economy*. Basingstoke, UK: Palgrave Macmillan.

Van den Berg, L., van der Meer, J., & Otgaar, A. H. J. (2007). The attractive city: Catalyst of sustainable urban development. In *XVI Congreso de Estudios Vascos* (pp. 485–491). Bilbao, Spain.

Vlacheas, P., Giaffreda, R., Stavroulaki, V., Kelaidonis, D., Foteinos, V., Poulios, G., et al. (2013). Enabling smart cities through a cognitive management framework for the internet of things. *IEEE Communications Magazine, 51*(6), 102–111.

Weber, A. S. (2011). The role of education in knowledge economies in developing countries. *Procedia—Social and Behavioral Sciences* (Vol. 15, pp. 2589–2594). Amsterdam, NL: Elsevier.

Weber, A. S. (2014). Education, development and sustainability in Qatar: A case study of economic and knowledge transformation in the Arabian Gulf. In A. W. Wiseman, N. H. Alromi, & S. Alshumrani (Eds.), *Education for a knowledge society in Arabian Gulf countries* (pp. 253–276). Bradford, UK: Emerald.

Weber, M. (1921). Die Stadt. Eine soziologische Untersuchung. *Archiv für Sozialwissenschaft und Sozialpolitik, 47,* 621–772.

Wiedmann, F., Mirincheva, V., & Salama, A. M. (2013). Urban reconfiguration and revitalisation: Public mega projects in Doha's historic centre. *Open House International, 38*(4), 27–36.

Wiseman, A. W., Alromi, N. H., & Alshumrani, S. (2014). Challenges to creating an Arabian Gulf knowledge economy. In A. W. Wiseman, N. H. Alromi, & S. Alshumrani (Eds.), *Education for a knowledge society in Arabian Gulf countries* (pp. 1–33). Bingley, UK: Emerald.

Wilkins, S. (2011). Who benefits from foreign universities in the Arab gulf states? *Australian Universities' Review, 53,* 73–83.

Witte, S. (2010). Gulf State branch campuses: Global student recruitment. *International Higher Education, 58,* 5–6.

Chapter 2
Qatar in a Nutshell

Abstract Qatar—one of the countries in the Gulf region—has developed a vision to transform its society into a knowledge-based society and is already in the process of implementing its plans. We introduce Qatar and its national vision as well as some peculiarities of the country. Basic aspects like Qatar's geography, climate, demography, language, culture, religion, politics, economy, public health and education are summarized.

Keywords Qatar · Knowledge society · Qatar National Vision 2030
Geography · Climate · Demography · Language · Culture · Religion
Politics · Rentier state · Rentier mentality · Economy · Kafāla
Public health · Education · Tradition · Modernity

2.1 Qatar as a Post-oil Knowledge-Based State?

In terms of Qatar's National Vision 2030, the security of the peoples' standard of living is to the fore, but the government realized that this cannot be the main and only goal of a country. With the focus on its values, Qatar has to balance some further challenges: modernizing traditions while ensuring their preservation, serving the needs of this and future generations, managing the country's growth and expansion, controlling the targeted size and the quality of the expatriate labor, going a path of development that is compatible to it, managing the economic growth and the social development and protecting the environment. To resolve these challenges and to fulfill the transformation of Qatar into an advanced country until 2030, the government outlines and promotes four pillars that the Vision rests on. On those pillars, the government has to work collectively with the Qatari people:

- **Human Development**
  Development of all its people to enable them to sustain a prosperous society.
- **Social Development**
  Development of a just and caring society based on high moral standards, and capable of playing a significant role in the global partnership for development.

J. Gremm et al., *Transitioning Towards a Knowledge Society*,
https://doi.org/10.1007/978-3-319-71195-9_2

- **Economic Development**
  Development of a competitive and diversified economy capable of meeting the needs of, and securing a high standard of living for, all its people for the present and for the future.
- **Environmental Development**
  Management of the environment such that there is harmony between economic growth, social development and environmental protection (GSDP, 2008, p. 11).

The General Secretariat for Development Planning (GSDP) will coordinate the tasks carried out by the implementation of these pillars. The National Vision becomes hence the framework of Qatar National Development Strategy (QNDS) 2011–2016. All plans of the strategy will be put into practice in cooperation with stakeholders, the private sector as well as ministries and government agencies, but can only be achieved with the support of the whole nation. The strategy's aim is

> to transform Qatar into an advanced country, sustaining its development and providing a high standard of living for all its people – for generations to come. It foresees a vibrant and prosperous Qatar with economic and social justice for all. It envisages all Qataris working together in pursuing these aspirations, with strong Islamic and family values guiding their collective energies (GSDP, 2011, p. 2).

To achieve this transformation, QNDS tries to put the Vision's plans into action, presents new initiatives and builds on what currently exists. It provides new impetus for those projects, policies and institutions that are brought into being by support of the National Vision. For the Qatar National Development Strategy main goals are to build a knowledge economy as well as to foster educating and training Qataris in order to their active participation in the knowledge society (GSDP, 2011, p. 122).

Guided by both—Qatar's National Vision and Qatar's National Development Strategy—Qatar Foundation (QF), a non-profit organization that supports the development of the knowledge economy by establishing many different programs and cutting-edge facilities in education, research and community development, published the Qatar Foundation Strategic Plan 2013–2030. Therewith, the organization translates the national needs—formulated in the four pillars of the National Vision—into five strategic objectives and adds one objective that they want to achieve internally as the organization itself to support QF's core mission activities:

1. Develop Human Capital
2. Drive Innovation to Advance Economic Diversification
3. Contribute Towards National Priorities
4. Foster a Progressive and Engaged Society Appreciative of its Heritage
5. Promote Productive Change Regionally and Internationally
6. Enhance Internal Capabilities, Sustainability and Synergy through Shared Services (QF, 2013, p. 1).

Characterized as a living document, the Strategic Plan will be reviewed and updated periodically. The core of all these initiatives is "to invest in the country's people as

Fig. 2.1 A big shell is reminiscent of the previous pearl diving era

they are the foundation of Qatar's future knowledge-based economy (KBE)" (QF, 2013, p. 2).

One of the states situated on the Arabian Peninsula is the state of Qatar—a sovereign and independent nation, ruled by His Highness Sheikh Tamim bin Hamad bin Khalifa Al Thani (Hukoomi, 2016a). Before its complete independence in 1971, Qatar was a poor British protectorate that was solely known for its pearl fishery (Commins, 2012; CIA, 2016; Fig. 2.1).

Made rich by large reserves of oil and gas, the country is more than capable to react to the thread of ending resources. Qatar has developed a plan to diversify its economy manifested in the Qatar National Vision 2030. The goal is to transform Qatar into a pioneering state with a knowledge-based and extremely competitive economy while maintaining its strong cultural and traditional values. Results are already visible in the progress that the country has made in economic, social and political fields (GSDP, 2008). What had been a small village in the past, was transformed into "a vibrant emerging regional urban centre with more than 1.7 million inhabitants" (Salama & Wiedmann, 2013, p. 61) during the second half of the last century. Before this transformation life in Doha, the capital city of Qatar, took place on the market, the port as well as in and around the mosques. Nowadays, we can find a spacious area with a new international airport, the artificial island The Pearl, the reconstructed market Souq Waqif (Fig. 2.2), the West Bay area (Fig. 2.3), the cultural village Katara where Arabic art and cultural facilities are promoted (Fig. 2.4), the 250 hectare large Aspire Zone that acts as the region's sports and shopping center and the also newly-built knowledge-intensive area Education City (Salama & Wiedmann, 2013) and Qatar University.

Furthermore, the prospect of Doha hosting the FIFA World Cup in 2022 has led to changes within the country. In respect of this spectacle, Qatar invests heavily in

Fig. 2.2 Souq Waqif

Fig. 2.3 Skyline of Doha's West Bay

Fig. 2.4 Katara cultural village

world-class infrastructure and in the improvement of its education and healthcare systems (CIA, 2016). By hosting this renowned sports event, Qatar has successfully managed to attract global interest and awareness as well as the marketing of Qatar as a brand. It is safe to say that Doha became “an important emerging regional and global capital in the Middle East” (Salama & Wiedmann, 2013, p. xxi) that attracts attention not only to the oil and gas sector, but also to culture and sports, tourism, politics and policy initiation.

2.2 Geography and Climate

Qatar is a small state on the Arabian Peninsula that borders on the Gulf (563 km long coastline) and on Saudi Arabia in the south (87 km long border). Its location in the center of the Gulf region and near the major petroleum deposits is strategically good. With 11,586 square kilometers, the country is almost twice the size of the U.S. state Delaware and just a little smaller than the U.S. state Connecticut. Qatar’s capital Doha is situated on the east coast of the country. On the whole, the terrain is almost flat and the desert is barren (CIA, 2016; Fig. 2.5).

The climate can be described as arid—winters are mild and very pleasant, summers are in contrast very hot and humid (CIA, 2016). The mean monthly temperature varies between 17 °C in January and 36 °C in July. In the summer, temperatures above 40 degrees are not uncommon. If it rains (which happens infrequently and mostly in winter months), the rain falls in brief showers (QTA, 2015). Haze, dust- and sandstorms are common occurrences (CIA, 2016).

Fig. 2.5 Qatar's location in the Gulf region. *Source* Google Maps

There are only few renewable natural water resources in Qatar, which are used mostly for agriculture and private consumption (Mohtar, 2016). Qatar's freshwater deficit is addressed by desalinated sea water. Desalination plants have high energy demand. However, "Qatar's domestic water use is among the highest in the world" (Mohtar, 2016, p. 297).

2.3 Demography and Language

The total population of Qatar amounts to 2,559,267 in April 2016 (MDPS, 2016a). The State is home to people from all over the world resulting in the occurrence of many different ethnic groups. Over the years, Qatari nationals became a minority in their own country with just 12 percent of the total population (Snoj, 2013). In line with this, while the official language of Qatar is Arabic, English is the primary language used for communication in the country (QSA, 2009).

Is there trust between the different groups living in Qatar? The mean values for general trust are lowest for Qatari (0.20) and white-collar Arab expats (0.23). "White-collar immigrants from Asia and migrant laborers—who are predominantly from nations outside of the Middle East—display much higher levels of generalized trust" (Diob et al., 2017, p. 193). Mean values of trust for white-collar Asians is 0.49 and for migrant workers it is 0.46. For Qatari, women are more trusting than men. The high values for trust of expats (outside of Middle East) are in line with the observation that some foreigners in Qatar "develop emotional connections to the states where they reside" (Koch, 2016, p. 52). When it comes to cyberbullying behavior among young people in Qatar (Foody, Samara, El Asam, Morsi, & Khattab, 2017), Onsor, Alameer, Almonabih, Alqahtani, and Tweem (2012) found that local children were more likely to be active bullies (21.1% in contrast to 17.4% for expat children) and non-nationals were more likely to be victims of bullying (39.9 vs. 33.1%).

Our examples of general trust and of cyberbullying exhibit that there are differences between the groups of inhabitants living in Qatar. The social coexistence of the groups, mainly (1) Qatari, (2) foreign white-collar expats, and (3) migrant workers, seems not to be easy and comfortable (Gardner et al., 2013; Nagy, 2006; Mohammad & Sidaway, 2016).

2.4 Culture and Religion

According to the constitution (Art. 1), Qatar's "religion is Islam and Shari'a law shall be a main source of its legislations". Key point in that formulation is that Shari'a is *a* source, and not the *one* source (Tok, Alkhater, & Pal, 2016, p. 8). Moslem religion is a source of the state of Qatar, but not the only one.

The National Vision of Qatar "seeks to achieve development progress, yet at the same time it aims to preserve national culture" (GSDP, 2008, p. 35). The traditional Muslim country Qatar presents itself as a society whose daily life is divided into two parts—on the one hand the life of the women and on the other hand the life of the men. Mostly everything takes place separately: classes are gender-segregated at Qatar University, men and women pray in different rooms, bride and groom celebrate their wedding parties apart from each other. As it seems for Qataris it is perfectly normal—for Westerns it is elusive (Lundsgaard Ottsen & Berntsen, 2013). Qatar is therefore characterized as a very traditional country with a clear majority of 77.5 percent being Muslims. In addition, 8.5 percent of the population are Christians and the remaining 14.0 percent are Hindus or of other Indian Religions (CIA, 2016). As representatives for the Arabian culture, living and preferences, Doha will run four impressive museums in the future that present a huge artistic range "[f]rom traditional Islamic craft and historical objects to contemporary art" (Qatar Museums, 2016, 1st paragraph). The Museum of Islamic Art (MIA) provides a complete overview of Islamic art (Fig. 2.6), the Arab Museum of Modern Art is particularly intended for the local community that is interested in the creativity of today's Arab world. In the foreseeable future the National Museum and the 3-2-1 Qatar Olympic and Sports Museum will make a big contribution to the country's cultural landscape. They will open to give visitors the chance to delve into Qatar's history, cultural heritage, further development projects and sporting spirit (Qatar Museums, 2016a). The current Emir's sister Sheikha Al Mayassa bint Hamad Al Thani chairs Qatar Museums (QM), which manages Qatar's art collections and museums.

The Katara Cultural Village—or more often known just as Katara—is a further cultural highlight in the north of Doha directly by the sea. The ninety-nine hectare big area is the most frequently visited urban space in Qatar. It is designed "to simulate a traditional, real or imagined, Qatari village with covered alleyways and

Fig. 2.6 Museum of islamic art

narrow pedestrian streets" (Salama, Khalfani, & Al-Maimani, 2013, p. 50) and houses among others different theatres, art galleries, souks, cafes and restaurants (Fig. 2.7).

Highlights are the amphitheater and the public access to the beach—but only accessible by paying admission money—that invite to stay (Wiedmann & Salama, 2013; Fig. 2.8).

"However, the village itself is designated for an exclusive clientele and patronage as it is gated; visitors are generally screened before being admitted" (Salama, Khalfani, & Al-Maimani, 2013, p. 50). Cultural or public space has become commercialized over time so that the use is not achievable for every one of every social class (Wiedmann & Salama, 2013).

Furthermore, the redesign of the old Souq Waqif illustrates the Arabian culture in a nice mixture of modern and traditional national elements in the old city center

Fig. 2.7 Traditional Qatari buildings in Katara

Fig. 2.8 Amphitheater in Katara cultural village

Fig. 2.9 Traditional goods at Souq Waqif

of Doha (Fig. 2.9). With the help of traditional construction techniques, the market is rebuilt as similar to the original as possible, in contrast to the new souk located directly by the sea. Surrounded by many shops, restaurants and cafes, the recreated shopping and meeting place is “a unique experiment to combine cultural heritage, tourist attraction, public realm and leisure space with the still remaining function as traditional market place” (Wiedmann, Mirincheva, & Salama, 2013, p. 30).

2.5 Politics

Qatar is a monarchy, which is ruled by the tribally based elite Al Thani family, originating from the Ma’adid clan of the Banu Tamim tribe (Gause, 1994) and whose governmental system is based on the separation of powers (Hukoomi, 2016b). The Emir is the Head of State of Qatar (Hukoomi, 2016a). In 2013, the former Emir Sheikh Hamad bin Khalifa Al Thani, who “has embarked on an ambitious program of modernizing the country” (Kamrava, 2009, p. 401), passed the power to his son Sheikh Tamim bin Hamad bin Khalifa Al Thani. This was the first time in Qatar’s history that a leader voluntarily retired from the position of Emir. The change in political leaders did not alter the structure of high-level decision making, but “what has differed since the 2011–2013 period is the larger regional and international contexts within which Qatari domestic and international policies are formulated, executed, and, importantly, perceived” (Kamrava, 2015, p. xii). Together with the legal heir the Emir constitutes the executive authority of Qatar. The legislative authority is vested in the Advisory Council and the judicial authority in courts of law. Set under the authority of the Emir, the Defense Council supervises the armed forces (Hukoomi, 2016b). A new permanent Constitution became effective on June 8, 2014 (Hukoomi, 2016c). As determined by this Constitution, the Council of Ministers supports the executive authority. It consists of the Prime Minister and six supreme councils. The Prime Minister is the head of the Council of Ministers and is appointed by the Emir just like all other ministers (Hukoomi, 2016b). Taking a deeper look at the monarchy one might find resemblances to the way companies are operated. The oil industry, which is owned by the government, is responsible for the power possessed by the state. Consequently, the decisions on how to spend oil revenues and who will profit from it and by how much is up to the rulers of Qatar. Thus, with oil and gas becoming the driving force of economy, the corporate identity and power of the ruling family of Qatar grows. Citizens are employed by the political elite and receive financial benefits from them. Because of that the citizens are tied to the rulers and are compelled to approve of the way of governance if they aim at success (Gause, 1994).

In terms of foreign and security politics, Qatar pursues the strategy of partnership with the United States of America. The U.S. run two large military bases in Qatar, namely Al Udeid air base and As Sayliyah base housing the army (Kamrava, 2015, p. 89). Al Udeid (near Doha) is the largest U.S. base in the Middle East accommodating about 10,000 troops.

Referring to Ulrichsen (2014), Qatar played an important role during the time of the Arab Spring. Qatar's Al Jazeera shaped the emerging narratives of the protest and mobilized Arab support for the protest movements. "At a time of great regional uncertainty, Qatar presented a compelling image as an outpost of stability and prosperity, even as the protests reached neighboring Gulf Cooperation Council (GCC) states" (Ulrichsen, 2014, p. 2).

2.6 Rentier State and Rentier Mentality

Due to the gas-based prosperity, there is no need for taxation in Qatar. Most of the government's revenues come from the international economy and the state does not have to collect any further money from its citizens. States with this character, where governments are the most powerful players in the local economy, are called "rentier states".

In a rentier state, "only few are engaged in the generation of this rent (wealth), the majority being only involved in the distribution and utilization of it" (Beblawi, 1987, p. 51). The principal recipient of the external rent is the state's government, which in turn distributes the rent to its citizens in the GCC states (but not, for instance, in Venezuela). This leads to a specific mentality of the citizens in the Arab oil-producing countries: the "rentier mentality". In a rentier state, there is a break in the work—reward causation. "Reward—income or wealth—is not related to work and risk bearing, rather to chance or situation", Beblawi (1987, p. 52) emphasizes. There are some implications of this rentier mentality.

- It is problematic to motivate the privileged citizens to work hard (as money comes anyway);
- if (non-privileged) expatriates take up employment in the rentier state, they have to do it under ward of a privileged citizen (leading to the *kafāla* system; Beblawi, 1987, p. 56); and
- the privileged citizens tend to minimize the number of privileged people (leading to massive problems for expatriates to acquire citizenship in the rentier state).

There is no doubt that Qatar nowadays is indeed a rentier state.

Here, one question is far more important than that of taxation: How to spend and distribute the "rent" revenues received by the gas business? This question can be answered as follows: by providing generous benefits to the citizens in the form of free education, healthcare, housing, consumer goods and services. "All this amounts to an absence of poor Qataris" (Kamrava, 2015, p. 112). This results in the contentment of the people and gives governments the freedom to exercise their power because there is nearly no opposition of the citizens. Citizens adopt the rentier mentality as they appreciate the benefits and also the jobs they receive from the state and do not press for political participation (Gause, 1994; Reiche, 2010). "Qatar's ruling elite, with their immense access to external rents and a small citizen population, has been better placed than many of their neighbors to engage extensively in distributional politics" (Babar, 2014, p. 409).

2.7 Economy

Qatar's pre-oil economy was dominated by the pearl industry; however, in the 1930s, demand for pearls collapsed due to the introduction of cultured pearls (Ulrichsen, 2016, p. 23). Production of oil in the Gulf area began in 1933 in Bahrain; in 1939, oil reserves in Qatar were discovered. Qatar's economy can be characterized as "state capitalism" (Ulrichsen, 2016, p. 68), as nearly all important oil and gas producing companies (as, for instance, Qatargas and Qatar Petroleum) as well as other non-oil/gas-based firms as Qatar Airways (transportation), Industries Qatar (heavy industry), Qatari Diar (real estate) or Ooredoo (telecommunication) are entirely state-owned (Ulrichsen, 2016, p. 71).

The oil and natural gas sector of Qatar is responsible for over 50 percent of the Gross Domestic Product (GDP) in 2016 (The Heritage Foundation, 2016). The country's reserves of gas are the third largest in the world (CIA, 2016) and in 2007, the arising businesses made Qatar even to the top liquefied natural gas (LNG) exporter of the world (Alraouf, 2012). Sourcing from Qatar's big North Field near Ras Laffan, Qatar Petroleum (Alfadala & El-Halwagi, 2017) runs 5 LNG production plants, each with a capacity of 7.8 million tons annually. Qatar Petroleum's 56 ships bring the produced LNG from Ras Laffan Port to the target markets (Esen & Oral, 2016). "A significant portion of the LNG shipped by Qatar is re-gasified in receiving countries where Qatar either co-owns terminals (Italy, UK and USA) or leases terminal capacity on a long-term basis (Belgium)" (Ibrahim & Harrigan, 2012, p. 9). Qatar's exports are dominated by gas and oil: in 2015, gas exports count for 64.8% of all exports, oil exports for further 13.6% and other petrochemical products for 4.4% (GTAI, 2016, p. 3). Main export countries include Japan (20.8% of all exports), South Korea (17.3%) and India (11.9%) (GTAI, 2016, p. 4). The huge gas reserves are the main economic strength of Qatar; however, they are also its major weakness, as Qatar is dependent on gas exports and the international gas prizes (Wright, 2017). "Most risks" of Qatar's economic future "are grounded in international oil-price movements", Qatar's Ministry of Development Planning and Statistics states (MDPS, 2015a, p. 2).

Due to its oil and gas resources, Qatar is not only the country with one of the world's highest per capita income, but also the country with the lowest unemployment rate. However, the oil prices have started to decrease remarkably leaving Qatar no choice but to diversify its economy (CIA, 2016). Consequently, the country's investment in non-energy sectors is gaining ground (Hukoomi, 2016a). The Ministry of Development Planning and Statistics "sees steady expansion in the non-oil and gas sector, with double-digit growth in 2015, but this will taper in 2016 and 2017 as investment and construction activity begins to plateau, and as population growth eventually slows" (MDPS, 2015b, p. iii). As stated in the country's National Vision 2030, Qatar aims at a

> diversified economy that gradually reduces its dependence on hydrocarbon industries, enhances the role of the private sector and maintains its competitiveness through:
>
> - Expansion of industries and services with competitive advantages derived from hydrocarbon industries
> - Design and development of economic activities in which Qatar can specialize, including the technical and human requirements of these activities
> - A knowledge-based economy characterized by innovation; entrepreneurship; excellence in education; a world-class infrastructural backbone; the efficient delivery of public services; and transparent and accountable government (GSDP, 2008, p. 29).

To strengthen this "post-carbon economy", Qatar established the Qatar Science and Technology Park (QSTP), a Free Zone (QSTP, 2016c), to attract "companies and institutes from around the world to develop and commercialize their technology in Qatar, and [to] help […] entrepreneurs launch technology businesses" (QSTP, 2016a, 1st paragraph). Free zones or free trade zones are an essential tool to attract international companies, entrepreneurs and investors. They establish businesses in the country, which grants benefits like partially or fully abolished taxation, the provision of modern and high-class infrastructure, the reduction of bureaucratic demands and more relaxed labor legislation (Salama & Wiedmann, 2013). Among other things, companies and institutes of Qatar Science and Technology Park benefit from tax exemption and duty-free import of needed goods and services (QSTP, 2016b). Qatar's plan to create a dynamic, competitive and more diversified economy is also manifested in its National Vision 2030, where the economic development is one of the four pillars of the vision (GSDP, 2008). Many companies

Fig. 2.10 Qatar Petroleum headquarters in the West Bay area

in Qatar, especially the powerful ones, strive to support this vision. One of them is Qatar Petroleum (QP), a state-owned public corporation which is responsible for all phases of the oil and gas industry (QP, 2015; Fig. 2.10). Its revenue in 2014 was about 169 billion QR (46 billion U.S. dollar) and its net income amounted to 113 billion QR (31 billion U.S. dollar) (QP, 2014).

Qatar Airways also contributes to the vision by shaping Doha as a key aviation hub—regional as well as global (Qatar Airways, 2016b). The airline is one of the most renowned airlines across the globe while growing faster than any other airline in the world. More than 40,000 professionals serve in this business. Just as in 2011 and 2012, Qatar Airways has been rewarded with the title "Airline of the year" by Skytrax in 2015 (Qatar Airways, 2016a). Qatar's path to a knowledge society is also supported by the global communications company Ooredoo. Its headquarters are based in Qatar, but it operates networks not only in the Middle East, but in North Africa, Southeast Asia and Subcontinent as well (Ooredoo, 2015). The TV news channel Al Jazeera, which employs journalists from more than 60 countries, has its headquarters in Qatar, too. Today, the channel is received by 220 million households in more than 100 countries (Hossili, 2013). Last but not least, Qatar owns some successful banks, like Qatar National Bank which is the biggest bank in Qatar and Qatar's leading Islamic Bank Qatar Islamic Bank, both with the ambition to expand their international presences (Qatar Islamic Bank, 2016; Qatar National Bank, 2016).

Due to the small population that lacks the necessary skills to drive Qatar's economic growth, the country's development depends largely on foreign workers (Moini, Bikson, Neu, & DeSisto, 2009). "The import of talent spans the full labor market" (Powell, 2014, p. 258). Those expatriate workers are controlled by the country's sponsor system called *kafāla* (Diop, Johnston, & Le, 2015). "The *kafāla* links individual guest workers to a particular job and, more importantly, to a particular citizen-sponsor or corporate entity (a *kafeel*)" (Gardner, 2010, p. 203). Entering one of the GCC states in the function of a guest worker is only possible through the *kafāla*. The guest worker signs a contract to work in the GCC for usually two years (Gardner, 2010). Till 2016, they were not allowed to change their employee; however, they may leave the country for holiday. Labor unions are banned.

Qatar-based firms are linked to companies all over the world. Salama, Wiedmann, Thierstein and Al Ghatam (2016) studied city connectivity from the office networks of multi-location advanced producer services and of high-tech firms. High-tech companies operate in the industries of chemistry and pharmacy, machinery, electronics, computer hardware, etc., while advanced producer services can be found in banking and finance, advertising and media, information and communication services, etc. (Salama et al. 2016, p. 287). The intra-firm connectivity of Qatar-based advanced producer services is greatest with New York, followed by London, Hong Kong, Tokyo, Paris and Singapore. However, the connectivity of high-tech companies is dominated by firms in Singapore, followed by companies in Moscow, Paris, Sao Paulo, Buenos Aires and Shanghai. "In contrast to Advanced Producer Service sector, High-Tech seems to be networked much more with Asian locations while North-American locations play no role" (Salama et al., 2016, p. 289).

Due to the increasing dependence on foreign workers, the government introduced a plan to counteract this phenomenon. The so-called "Qatarization", a concept that requires that a certain percentage of Qataris have to be employed in private-sector companies, was introduced by the Qatari Labor Law of 1962 and mentioned again in 2004 (Zakhidov, 2015). This strategy was implemented by the country to bring Qataris into professional and managerial jobs especially in the private sector which are currently occupied by expatriate workers. The government focusses strongly on the private sector because only a really few percent of all employed nationals work for the truly private sector (Moini et al., 2009; Powell, 2014).

2.8 Public Health

With the discovery of natural resources and the wealth that resulted from it, Qatar embraced the opportunity to provide social services, like healthcare (Ali, Gjebrea, Sifton, Alkuwari, & Atun, 2016), to its citizens free of charge, which were more or less neglected by the government before (Gause, 1994). In 2013, health expenditures amounted to 2.2 percent of the country's GDP. Especially the new Emir prioritizes the expansion of Qatar's healthcare system and makes sure to establish high-class infrastructures in this regard (CIA, 2016). The improvement of healthcare in the country is also declared as a long term goal in Qatar's National Vision 2030. The idea of GSDP (2008, p. 17) is to establish a

> comprehensive world-class healthcare system whose services are accessible to the whole population, including:
>
> - Effective and affordable services in accordance with the principle of partnership in bearing the costs of health care
> - Coverage of preventive and curative health care, both physical and mental, taking into account the differing needs of men, women and children
> - High quality research directed at improving the effectiveness and quality of healthcare
>
> An integrated system of health care offering high-quality services through public and private institutions operating under the direction of a national health policy that sets and monitors standards for social, economic, administrative and technical aspects of health care.
>
> A skilled national workforce capable of providing high quality health services.
>
> Continued commitment by the state to provide sufficient funds for maintaining the health of Qatar's population in accordance with the principle of partnership in bearing the costs of health care.

Additionally, Sheikha Moza bint Nasser Al Missned, wife of the former Emir, launched Qatar's National Health Strategy 2011–2016 in 2011 to support this vision and to move Qatar towards the achievement of these objectives (SCH, 2013).

Hamad Medical Corporation (HMC) is not only Qatar's main provider of secondary and tertiary healthcare, but also one of the leading hospital providers in the Middle East. It already manages eight hospitals and plans an expansion in the form of four additional hospitals to support Qatar's plan for the future (Hamad Medical Corporation, 2016). All in all, in 2015 there are 15 hospitals with about 3000 beds in Qatar.

Also Sidra Medical and Research Center, which is a hospital for women and children, shall support Qatar's vision in terms of healthcare (Fig. 2.11). For this purpose, they seek to collaborate with Weill-Cornell Medicine, Hamad Medical Corporation and Qatar Biobank. The medical center is a member of the Qatar Foundation for Education, Science and Community Development. On May 1, 2016 it has opened its first Outpatient Clinic and is planning to introduce further clinics and services as the year progresses (Sidra, 2016).

The Orthopaedic and Sports Medicine Hospital Aspetar located in Qatar is the first hospital in the Gulf region that provides medical treatment for sports-related injuries. Some of the best practitioners and researchers in the field of sports medicine are employed in the ultra-modern facility located at the Aspire Zone (Aspetar, 2015).

Fig. 2.11 Sidra Medical and Research Center. *Photo* Tamara Heck (2016, personal communication)

2.9 Education

According to the National Vision 2030, one of the most important developments in Qatar is that of a

> world-class educational system that equips citizens to achieve their aspirations and to meet the needs of Qatar's society, including:
>
> - Educational curricula and training programs responding to the current and future needs of the labor market
> - High quality educational and training opportunities appropriate to each individual's aspirations and abilities
> - Accessible educational programs for life-long learning (GSDP, 2008, p. 16)

To advance the education system in Qatar, an Emiri Decree established the Qatar Foundation for Education, Science and Community Development in 1995. Under the leadership of Sheikha Moza the private, non-profit organization has set itself to improve the quality of life in Qatar and the surrounding region by providing world-class educational opportunities. This vision is realized in a new higher education campus located in the northwest of Doha called Education City, which is, according to Khodr and Reiche (2012, p. 149), "a city in-the-city". Based on the model of the branch campus concept, the 2500 acres big area accommodates world-class universities that "offer their most prestigious programmes to Qatar as fully-fledged partners of Qatar Foundation" (Salama & Wiedmann, 2013, p. 55). These foreign universities that offer their programs to Qatar's students are mostly from the UK, the United States and France and include, according to Crist (2015): Virginia Commonwealth University (Art and Design), Weill Cornell Medical College (Healthcare), Northwestern University (Communication, Journalism), HEC Paris (Business and Management), and University College London (Librarianship and Museum Practice). The world's most expensive national education project is the first step of Doha's plan to become an important knowledge center with a knowledge-based economy (Figs. 2.12 and 2.13). Since 2016, Sheikha Moza's daughter and sister of the current Emir Sheikha Hind bint Hamad Al-Thani is CEO and vice-chairperson of Qatar Foundation.

Already well before its establishment in 1973, Qatar University (QU)—the "country's national and major institution of higher education" (QU, 2016, 1st paragraph)—was founded. "[A]cross all Arab countries and especially the Arab Gulf region, higher education, indeed education generally, has been extended substantially in the past 2 decades" (Donn & Al Manthri, 2010, p. 97). In 2003, the RAND-Qatar Policy Institute (RQPI) was engaged to support the reform to strengthen Qatar University (Fig. 2.14) to adhere to the needs of the younger generations responsible for Qatar's future in connection with the establishment of a knowledge society (Moini et al., 2009).

Nowadays, the university has reached substantial international attention and is frequently mentioned in the international media. According to Times Higher Education, Qatar University is the most international university in the world. This is the result of the 2015–2016 ranking of 800 universities considered altogether. The

Fig. 2.12 Carnegie Mellon University in Qatar as an example for the prestigious architecture in Education City

Fig. 2.13 Ceremonial court in Education City

Fig. 2.14 Qatar University

indicator used for this ranking is the international outlook indicator of the Times Higher Education World University Rankings 2015–2016. This indicator includes the proportion of international staff and students and of research papers published with at least one international co-author. The overall ranking examined, beside the international outlook indicator, four further indicators: teaching, research, citations and industry income. It is striking that while Qatar University comes first in the International University Ranking, it ends up on one of the last ranks (601-800) in the overall ranking of 800 universities (Bothwell, 2016). Different colleges from Islamic studies via engineering through to medicine and pharmacy, a central library, an information technology center, as well as sports and recreational facilities are offered at Qatar University (Salama & Wiedmann, 2013; QU, 2016d).

Due to low scores in the Trends in International Mathematics and Science Study (TIMSS) and the Program for International Assessment (PISA), Qatar tasks the RAND Corporation, a non-profit organization, to rate the K-12 education system and build a new strategy from the bottom (Ellili-Cherif, Romanowski, & Nasser, 2012). The project started in 2002 and is called Education for a New Era (EFNE), an education that makes it possible to build "a world-class system that would meet the country's changing needs" (Brewer et al., 2007, p. xvii). Before that, there was no clear idea of how to provide quality education, the performance was low, school buildings were in poor conditions, classrooms were overcrowded and teachers were badly paid and had little training opportunities. Nowadays, the case has changed: The organization "has created immense opportunities for international educational consulting with numerous organizations flooding Qatar to offer their "expertise" in a variety of educational areas" (Ellili-Cherif et al., 2012, p. 471). Teachers are registered and have to acquire a license now, so that the Supreme Education Council (SEC) is able to manage, monitor and assess the teachers and their practice. Furthermore, all government schools were transformed into independent schools that can act autonomously now, the "curriculum standards that address both academic and non-academic dimensions of learning" were established and also the teaching strategies were changed "to be more student-centered, using modern information and communication technology, and conducting scientific research" (SEC, 2011, p. 7). The plans are supposed to be implemented until 2016. Time will tell how successful the results were until now, are today and will be in the future and to what extent education has accomplished its specific tasks of "ending illiteracy, diversification of hydrocarbon-based economies, and preservation of cultural integrity" (Weber, 2011, p. 2589). Said states for Qatar in 2016, that students' "science achievement on the TIMMS and PISA tests is significantly below the international average despite the progress achieved following education reforms and implementation of new curriculum standards" (Said, 2016, p. 2254). While 500 is the average international score, students at schools in Qatar perform in TIMMS in 2011 with 394 points (grade 4) and 419 points (2011). This is below the international average, but there is indeed progress, as grade 4 students had in 2007 only 294 points and grade 8 students 319 points (Said, 2016, p. 2258). The PISA achievements of students in Qatar are similar to the TIMMS results. In 2012, 15-year old students reached 384 points (again, 500 is the international average);

however, there is a clear improvement in contrast to 2006 (349 points) and more or less the same result as in 2009 (379) (Said, 2016, p. 2259).

Besides different public, school, academic, special and governmental libraries the indeed most important one is the new Qatar National Library (QNL) that is located in Education City (QNL, 2016b). Likewise, under the head of Qatar Foundation and therefore of Sheikha Moza, the library takes on the three tasks of being a "National Library, University and Research Library, and a Metropolitan Public Library of the digital age" (QNL, 2016a, 2nd paragraph). Today 1.2 million printed books, 500,000 ebooks, full-text databases, periodicals and newspapers, as well as special collections are offered. Everybody living in Qatar with a valid Qatari ID or residence permit already has access to a vast amount of online resources (QNL, 2016a). With the establishment of this project, Qatar made a further attempt to become a knowledge society, as

> the number of information centers is one of the indicators of a country's preparedness to support a knowledge-based economy. More information centers collecting data means more information and knowledge can reside in more places and thus more people can access them (Wand, 2016, p. 164).

2.10 Tradition Versus Modernity

With the decision of the political leaders to establish a knowledge society while maintaining local culture and traditions (GSDP, 2008), the state of Qatar has become a showplace of many different facets. The main location of this dichotomy is the capital city Doha: "Many parts of the city resemble a surreal and incongruent mixture of Hong Kong, on the one hand and Tucson, Arizona, on the other" (Kamrava, 2015, pp. 5–6). While giant skyscrapers with glittering facades and other high-end, ultramodern buildings have been raised on the one side, other parts are kept traditional or are rebuild by imitating the old, original style. The West Bay of Doha, located at the corniche, can be described as a futuristic neighborhood illuminated by millions of lights at nighttime and constantly under construction to become even more imposing (Fig. 2.15).

Right at the other side of the corniche, the old, traditional Souq Waqif is situated (Fig. 2.16), which is an Arabian marketplace where, especially in the evening, many people—mostly dressed in their national costumes—meet, socialize and purchase the offered goods.

However, this is not the only visual contrast in the country as there are further modern, but also traditional places in the interior. The Aspire area is a sports complex in Qatar which houses, among sports facilities, modern buildings including Villaggio—a large shopping mall with entertainment facilities, like an ice rink and an amusement park. Additionally, the artificial island called The Pearl is a highly attractive state-of-the-art complex of stores, cafes and apartments for Qatar's residents right by the sea (Fig. 2.17).

Fig. 2.15 Doha West Bay by night

Fig. 2.16 Birds, cats and dogs are for sale in the narrow alleys of Souq Waqif

Fig. 2.17 The artificial Island The Pearl

Admittedly, the sudden wealth has incrementally changed the city space by constructing skyscrapers, high-rise buildings and mansions—one more impressive than the other. However, at the same time parts of the poor, nomadic lifestyle that characterized the life of nationals before the oil boom have been maintained: Tents decorate the desert that serve as hideouts for Qataris during the weekends and a traditional animal market where sheep, goats and even camels are sold still exists in the country (Fig. 2.18).

These differences are, however, not only restricted to the cityscape as Qatar's people can also be described as divided in multiple ways. On the one hand, they are divided by their way of thinking as cultural attitudes are slowly changing in Qatar (Al-Kitbi, 2010). Some of the residents, among them Sheikha Al Mayassa Hamad bin Khalifa Al Thani, strive to adapt modern views and embrace and try new ways, while others strictly stick to the local culture and religion and do not accept anything new if they feel that it offends religious sensibilities. On the other hand, they are divided by their status in the country. In Qatar, there are three different groups of workers that need to be distinguished: Qataris, foreign talents and low-skilled foreign workers. Qataris who often lack education and skills traditionally work mainly in the easier and well-paid public sector, while well-educated expats nowadays fill professional and managerial positions in private companies and low-skilled expats perform poorly payed jobs in the construction and service sectors (Hertog, 2011; Moini et al., 2009). Especially with the prospect of hosting the FIFA

Fig. 2.18 Qatar animal market

World Cup in 2022, unskilled workers are needed for project-related work (Fig. 2.19) and thereby face conditions of modern slavery (Dorsey, 2015).

For Al-Ammari and Romanowski (2016, p. 1537) "globalization and modernization have benefitted Qatar. These benefits include a massive educational reform, access to world-class education (…), an increase in living standards, improved transportation, financial opportunities and changes in society and lifestyles". The Qatari family lives between the poles of the nuclear family, the tribe, the housekeeping personnel and the cultural changes (Kassem & Al-Muftah, 2016). The changes in culture affect polygamy; there is a decline since fewer women are willing to become second wife and first wives prefer divorce to accepting a second wife (Al-Ammari & Romanowski, 2016, p. 1537). About three quarter of Qatari respondents to Al-Ammaris's and Romanowski's survey agree, that it is better for a man to have only one wife. However, 7.9% of Qatari men have two wives, and further 0.7% are married with more than two women (2015; MDPS, 2016b, p. 12). The age of the first marriage is on average 23.8 years (women) and 26.3 years (men) (2015; MSPS, 2016b, p. 9). The marriage rate per 1000 Qatari population (15 years+) is 22.2 for females and 24.1 for males (2015; MDPS, 2016b, p. 6). Marriages are mainly arranged (62.3%) especially by parents and other relatives, but also by introducing the partner (20.2%) and by choice of the spouses themselves (17.5%). For most of the Qatari, the authority of the father should be respected under any circumstances (Al-Ammari & Romanowski, 2016, p. 1549). Due to the

Fig. 2.19 The stadiums for the FIFA World Cup 2022 (here in the Aspire Zone) are still under construction

wealth of the nation, household work is not popular among Qatari. Nearly 80% of Qatari households employ at least one domestic worker; about 6% have 5 or more employees for housekeeping (Al-Ammari & Romanowski, 2016, p. 1550).

One cannot overlook consanguinity (especially, marriage of first cousins) in Qatar. Consanguinity is a common marriage pattern within the Arab world; however, it could be a risk factor for health problems. Sandridge, Takeddin, Al-Kaabi, and Frances (2010) asked Qatari born between 1946 and 1991 on their relations to their spouses, for themselves and for their parents. 22% of the parent generation lived in a consanguineous relation and further 15% married a member of the same tribe. For the generation of the children, the values changed in favor of consanguinity. 35% of the respondents were married to a close relative and 9% to a member of the same tribe (Sandridge et al., 2010, p. 65). Two reasons dominate the justification of consanguinity. Those who married a close relative see benefits that (1) the bride or groom is already familiar to the partner and (2) that consanguinity promotes stability, traditions and continuity of culture and way of life (Sandridge et al., 2010, p. 68).

Prospectively, Qatar has to face the challenge "of aligning Arabian Gulf expectations, traditions, and norms with those of knowledge economies" (Wiseman, Alromi, & Alshumrani, 2014, p. 2).

We now have an overview of how Qatar is positioned in terms of economic activities, health conditions and the education system. To what extent do these and of course many other factors influence the development status of the country? In this context, we additionally want to introduce the concepts of knowledge economy and of knowledge-based development, followed by explanations of the country's strides towards these concepts in the subsequent chapters of this book.

References

Al-Ammari, B., & Romanowski, M. H. (2016). The impact of globalisation on society and culture in Qatar. *Pertanika Journal of Social Science and Humaniies, 24*(4), 1535–1556.

Alfadala, H. E., & El-Halwagi, M. M. (2017). Qatar's chemical industry: Monetizing natural gas. *Chemical Engineering Progress, 113*(2), 38–41.

Ali, F. M. H., Gjebrea, O., Sifton, C., Alkuwari, A., & Atun, R. (2016). Health policy making in a transformative state. In M. E. Tok, L. R. M. Alkhater, & L. A. Pal (Eds.), *Policy-making in a transformative state. The case of Qatar* (pp. 179–212). London, UK: Palgrave Macmillan.

Al-Kitbi, E. (2010). Where do GCC women stand in the development process? In C. Koch & L. Stenberg (Eds.), *The EU and the GCC: Challenges and prospects under the Swedish EU presidency* (pp. 95–103). Dubai, AE: Gulf Research Center.

Alraouf, A. A. (2012). Dohaization: Constructing a new urbanity brand amid Gulf and Middle Eastern cities. Retrieved from www.analchemyofarchitecture.blogspot.com.

Aspetar. (2015). *About Aspetar*. Retrieved from www.aspetar.com.

Babar, Z. R. (2014). The cost of belonging: Citizenship construction in the state of Qatar. *Middle East Journal, 68*(3), 403–420.

Beblawi, H. (1987). The rentier state in the Arab World. In H. Beblawi & G. Luciani (Eds.), *Nation, state and integration in the Arab world. Vol. 2: The Rentier State* (pp. 49–62). London, UK: Croom Helm, New York, NY: Methuen.

Bothwell, E. (2016, January 14). The world's most international universities 2016. *Times Higher Education*. Retrieved from www.timeshighereducation.com.

Brewer, D. J., Zellman, G. L., Ryan, G., Goldman, C. A., Stasz, C., & Constant, L. (2007). *Education for a New Era: Design and implementation of K–12 education reform in Qatar*. Santa Monica, CA: RAND Corporation.

CIA. (2016). *The World Factbook Qatar*. Retrieved from www.cia.gov.

Commins, D. (2012). *The Gulf states: A modern history*. London, UK: I.B. Tauris & Co Ltd.

Crist, J. T. (2015). Innovation in a small state: Qatar and the IBC cluster model of higher education. *The Muslim World, 105*(1), 93–115.

Diob, A., Jardina, A. E., Tessler, M., & Wittrock, J. (2017). Antecedents of trust among citizens and non-citizens in Qatar. *Journal of International Migration and Integration, 18*(1), 183–202.

Diob, A., Johnston, T., & Le, K. T. (2015). Reform of the *kafāla* system: A survey experiment from Qatar. *Journal of Arabian Studies. Arabia, the Gulf, and the Red Sea, 5*(2), 116–137.

Donn, G., & Al Manthri, Y. (2010). *Globalisation and higher education in the Arab Gulf states*. Oxford, UK: Symposium Books Ltd.

Dorsey, J. M. (2015). How Qatar is its own worst enemy. *The International Journal of the History of Sports, 32*(3), 422–439.

Ellili-Cherif, M., Romanowski, M. H., & Nasser, R. (2012). All that glitters is not gold: Challenges of teacher and school leader licensure licensing system in Qatar. *International Journal of Educational Development, 32,* 471–481.

Esen, V., & Oral, B. (2016). Natural gas reserve/production ratio in Russia, Iran, Qatar and Turkmenistan: A political and economic perspective. *Energy Policy, 93,* 101–109.

Foody, M., Samara, M., El Asam, A., Morsi, H., & Khattab, A. (2017). A review of cyberbullying legislation in Qatar: Considerations for policy makers and educators. *International Journal of Law and Psychiatry, 50,* 45–51.

Gardner, A. M. (2010). Engulfed: Indian guest workers, Bahraini citizens and the structural violence of the kafala system. In N. De Genova & N. Peutz (Eds.), *The deportation regime: Sovereignty, space, and the freedom of movement* (pp. 197–223). Durham, NC: Duke University Press.

Gardner, A. M., Pessoa, S., Diop, A., Al-Ghanim, K., Trung, K. L., & Harkness, L. (2013). A portrait of low-income migrants in contemporary Qatar. *Journal of Arabian Studies. Arabia, the Gulf, and the Red Sea, 3*(1), 1–17.

Gause, F. G. (1994). *Oil monarchies: Domestic and security challenges in the Arab Gulf states.* New York, NY: Council on Foreign Relations.

GSDP. (2008). *Qatar National Vision 2030.* Doha, QA: General Secretariat for Development Planning.

GSDP. (2011). *Qatar national development strategy 2011–2016.* Doha, QA: General Secretariat for Development and Planning.

GTAI. (2016). *Katar. Wirtschaftsdaten kompakt.* Bonn, Germany: Germany Trade & Invest.

Hamad Medical Corporation. (2016). *Our organization.* Retrieved from www.hamad.qa.

Hertog, S. (2011, May 31). The cost of counter-revolution in the GCC. *Foreign Policy.* Retrieved from www.foreignpolicy.com.

Hossili, P. (2013, December 18). Hello, this is Al Jazeera. Retrieved from www.hossli.com.

Hukoomi. (2016a). *About Qatar.* Retrieved from www.gov.qa.

Hukoomi. (2016b). *Government and legislatives.* Retrieved from www.gov.qa.

Hukoomi. (2016c). *The constitution.* Retrieved from www.gov.qa.

Ibrahim, I., & Harrigan, F. (2012). Qatar's economy: Past, present and future. *QScience Connect 2012, 9* (24 pp.).

Kamrava, M. (2009). Royal factionalism and political liberalization in Qatar. *Middle East Journal, 63*(3), 401–420.

Kamrava, M. (2015). *Qatar: Small state, big politics.* Ithaca, NY: Cornell University Press.

Kassem, L. M., & Al-Muftah, E. (2016). The Qatari family at the intersection of policies. In M. E. Tok, L. R. M. Alkhater, & L. A. Pal (Eds.), *Policy-making in a transformative state. The case of Qatar* (pp. 213–239). London, UK: Palgrave Macmillan.

Khodr, H., & Reiche, D. (2012). The specialised cities of the Gulf cooperation council: A case study of a distinct type of policy innovation and diffusion. *Digest of Middle East Studies, 21*(1), 149–177.

Koch, N. (2016). Is nationalism just for nationals? Civic nationalism for non citizens and celebrating National Say in Qatar and the UAE. *Political Geography, 54,* 43–53.

Lundsgaard Ottsen, C. L., & Berntsen, D. (2013). The cultural life script of Qatar and across cultures: Effects of gender and religion. *Memory, 22*(4), 390–407.

MDPS. (2015a). *Qatar economic outlook 2015–2017.* Doha, QA: Ministry of Development Planning and Statistics.

MDPS. (2015b). *Qatar economic outlook 2015–2017 Update.* Doha, QA: Ministry of Development Planning and Statistics.

MDPS. (2016a). *Monthly figures on total population.* Doha, QA: Ministry of Development Planning and Statistics.

MDPS. (2016b). *Marriage & divorce in the State of Qatar, 2015.* Doha, QA: Ministry of Development Planning and Statistics.

Mohammad, R., & Sidaway, J. D. (2016). Shards and stages. Migrant lives, power, and space viewed from Doha, Qatar. *Annals of the American Association of Geographers, 106*(6), 1397–1417.

Mohtar, R. H. (2016). Integrated water, energy, and food governance: A Qatari perspective. In M. E. Tok, L. R. M. Alkhater, & L. A. Pal (Eds.), *Policy-making in a transformative state. The case of Qatar* (pp. 295–397). London, UK: Palgrave Macmillan.

Moini, J. S., Bikson, T. K., Neu, C. R., & DeSisto, L. (2009). *The reform of Qatar University*. Santa Monica, CA: RAND Corporation.
Nagy, S. (2006). Making room for migrants, making sense of difference: Spatial and ideological expressions of social diversity in urban Qatar. *Urban Studies, 43*(1), 119–137.
Onsor, A., Alameer, A., Almonabih, H., Alqahtani, B., & Tweem, O. (2012). *Bullying amongst students: Characteristics, reasons and consequences*. Doha, QA: Al Aween Social Rehabilitation Center.
Ooredoo. (2015). *Annual report 2015*. Doha, QA: Ooredo.
Powell, J. J. (2014). University roots and branches between "Glocalisation" and "Mondialisation": Qatar's (Inter)national Universities. In A. W. Wiseman, N. H. Alromi, & S. Alshumrani (Eds.), *Education for a knowledge society in Arabian Gulf countries* (pp. 253–276). Bingley, UK: Emerald.
Qatar Airways. (2016a). *Group chief executive's message*. Retrieved from www.qatarairways.com.
Qatar Airways. (2016b). *The Qatar airways story*. Doha, QA: Qatar Airways.
Qatar Islamic Bank. (2016). *Our profile*. Retrieved from www.qib.com.qa.
Qatar Museums. (2016). *Museums & galleries*. Retrieved from www.qm.org.qa.
Qatar National Bank. (2016). *About QNB Group*. Retrieved from www.qnb.com.
QF. (2013). *Qatar foundation strategic plan 2013–2023*. Doha, QA: Qatar Foundation.
QNL. (2016a). *About the library*. Retrieved from www.qnl.qa.
QNL. (2016b). *The directory of libraries in Qatar*. Retrieved from www.qnl.qa.
QP. (2014). *1974–2014 40 years of excellence: Annual report 2014*. Doha, QA: Qatar Petroleum.
QP. (2015). *Qatar petroleum: Commited to excellence*. Retrieved from www.qp.com.qa.
QSA. (2009). *General information about Qatar*. Retrieved from www.qsa.gov.qa.
QSTP. (2016a). *About us*. Retrieved from www.qstp.org.qa.
QSTP. (2016b). *Facilities and services*. Retrieved from www.qstp.org.qa.
QSTP. (2016c). *Free zone*. Retrieved from www.qstp.org.qa.
QTA. (2015). *National profile—introduction to our country*. Retrieved from www.qatartourism.gov.qa.
QU. (2016). *Qatar University*. Retrieved from www.qu.edu.qa.
Reiche, D. (2010). Energy policies of Gulf Cooperation Council (GCC) countries: Possibilities and limitations of ecological modernization in rentier states. *Energy Policy, 38*, 2395–2403.
Said, Z. (2016). Science education reform in Qatar: Progress and challenges. *Eurasia Journal of Mathematics, Science & Technology Education, 12*(8), 2253–2265.
Salama, A. M., Khalfani, F., & Al-Maimani, A. (2013). Experiential assessment of urban open spaces in Doha. *Open House International, 38*(4), 47–57.
Salama, A. M., & Wiedmann, F. (2013). *Demystifying Doha*. Farnham, UK: Ashgate.
Salama, A. M., Wiedmann, F., Thierstein, A., & Al Ghatam, W. (2016). Knowledge economy as an initiator of sustainable urbanism in emerging metropolises: The case of Doha, Qatar. *ArchNet—International Journal of Architectural Research, 10*(1), 274–324.
Sandridge, A. L., Takeddin, J., Al-Kaabi, E., & Frances, Y. (2010). Consanguinity in Qatar: Knowledge, attitude and practice in a population born between 1946 and 1991. *Journal of Biosocial Science, 42*(1), 59–82.
SCH. (2013). *National Health strategy 2011–2016: Caring for the future*. Doha, QA: Supreme Council of Health.
SEC. (2011). *Education and training sector strategy 2011–2016 executive summary*. Doha, QA: Supreme Education Council.
Sidra. (2016). *Sidra Medical and Research Center*. Retrieved from www.sidra.org.
Snoj, J. (2013, December 13). Population of Qatar by nationality. *BQ magazine*. Retrieved from www.bq-magazine.com.
The Heritage Foundation. (2016). *Qatar*. Retrieved from www.heritage.org.
Tok, M. E., Alkhater, L. R. M., & Pal, L. A. (2016). Policy-making in a transformative state: The case of Qatar. In M. E. Tok, L. R. M. Alkhater, & L. A. Pal (Eds.), *Policy-making in a transformative state. The case of Qatar* (pp. 1–35). London, UK: Palgrave Macmillan.

Ulrichsen, K. C. (2014). *Qatar and the Arab spring. Policy drivers and regional implications*. Washington, DC: Carnegie Endowment for International Peace.

Ulrichsen, K. C. (2016). *The Gulf states in international political economy*. Basingstoke, UK: Palgrave Macmillan.

Wand, P. A. (2016). Correlating information centers to emerging knowledge-based economies. In A. Click, S. Ahmed, J. Hill, & J. Martin (Eds.), *Library and information science in the Middle East and North Africa* (pp. 156–172). Berlin, DE: De Gruyter.

Weber, A. S. (2011). The role of education in knowledge economies in developing countries. *Procedia-Social and Behavioral Sciences, 15,* 2589–2594.

Wiedmann, F., Mirincheva, V., & Salama, A. M. (2013). Urban reconfiguration and revitalisation: Public mega projects in Doha's historic centre. *Open House International, 38*(4), 27–36.

Wiedmann, F., & Salama, A. M. (2013). From pre-oil settlement to post-oil hub: The urban transformation of Doha. *International Journal of Architectural Research, 7*(2), 146–159.

Wiseman, A. W., Alromi, N. H., & Alshumrani, S. (2014). Challenges to creating an Arabian Gulf knowledge economy. In A. W. Wiseman, N. H. Alromi, & S. Alshumrani (Eds.), *Education for a knowledge society in Arabian Gulf countries* (pp. 1–33). Bingley, UK: Emerald.

Wright, S. (2017). Qatar's LNG: Impact of changing East-Asian market. *Middle East Policy, 24* (1), 154–165.

Zakhidov, D. (2015). K-12 education in Qatar. In K. C. Ulrichsen & A. Marnicio (Eds.), *The 2015 Doha conference: Findings from the student-led public diplomacy program* (pp. 9–13). Doha, QA: Rice University's Baker Institute for Public Policy.

Part II
Concept

Chapter 3
Knowledge Economy and Knowledge-Based Development

Abstract Knowledge economy and knowledge-based development are concepts that have emerged recently with the development of the knowledge society. Knowledge economy is a new kind of economy that is driven by knowledge and information. When knowledge is the driving force in all fragments of life, we talk about a knowledge society. To analyze Qatar's development status on its way to become a knowledge society, the Human Development Index, the ICT Development Index and the Networked Readiness Index are used. In comparison to the rest of the Gulf Corporation Council States, Qatar performs best in respect to these indicators.

Keywords Knowledge society · Knowledge economy · Knowledge-based development · Human development index (HDI) · ICT development index (IDI) Networked readiness index (NRI) · Qatar · GCC countries

3.1 Definition of the Concepts

Over time, society has experienced major transformations triggered by "rapid innovations in *the system of societal technology*" (Masuda, 1980, p. vii). The industrial society has superseded the agricultural society which has in turn superseded the hunting society. Nowadays knowledge has become the dominant factor that not only determines the economic sector—where finding employment is getting better and better for knowledge workers that arise from premium knowledge institutions (Castells, 2010; Florida, 2002)—but also governmental matters, education and health care (Ergazakis, Metaxiotis, & Psarras, 2006). With knowledge becoming the center of attention in the modern world the transition into an "information society" or "knowledge society" is already far advanced (Webster, 2006). The term *information society* represents thereby the technical perspective of a society where innovations are based on information and communication technologies (ICT). A knowledge society goes even further by being an information society in which the information content becomes a priority and knowledge will be available for everyone at any time and any place (Linde & Stock, 2011). Castells

J. Gremm et al., *Transitioning Towards a Knowledge Society*,
https://doi.org/10.1007/978-3-319-71195-9_3

(2010) refers to this new form of society as "network society" because of the rapid development of digital networking technologies that enable global interaction and knowledge exchange on a grand scale. The new kind of society "is organized around knowledge, for the purpose of social control and the directing of innovation and change" (Bell, 1999, p. 20). The importance of knowledge continues to rise unstoppably in this society (Castells, 1989), where "cities are taking a leading role as both knowledge-consumption and knowledge production hubs" (Carrillo, Yigitcanlar, García, & Lönnqvist, 2014, p. xvii). Since Qatar—our object of study—consists almost exclusively of its capital city Doha (El-Haddad, 2003), we will introduce a term for those cities: "informational cities"—as the prototypical city of a knowledge society (Castells, 1989). In these cities the flow of money, power and information are taken together as the "space of flows" that dominates the geographical "space of places". Considering the latest literature there is a more common term for this phenomenon, namely "smart city". The definition of smart city in the broader sense corresponds to our understanding of an informational city (Fietkiewicz & Stock, 2015). Giffinger et al. (2007) summarize in this term different factors like smart economy, smart people, smart governance, smart mobility, smart environment and smart living. In the contrary narrower sense, a smart city focuses on the environment, is a green city and is ICT-driven, but not based on it (Hall et al., 2000). "As society becomes increasingly knowledge-based, the nature of city development changes because activities in the knowledge sector are becoming more important" (Knight, 1995, p. 226). To support the establishment of knowledge societies, state- and citywide "knowledge-based development" (KBD) strategies need to be taken into consideration (Yigitcanlar & Velibeyoglu, 2008). Knowledge-based development describes the implementation of knowledge-intensive resources into local development (Knight, 1995) or into urban development when referring to the "urban specialized version, Knowledge-Based Urban Development (KBUD)" (Carrillo et al., 2014, p. xviii). The concept of knowledge-based development cannot only be applied to knowledge- and technology-intensive spaces (of high-end scientific, educational and innovation sectors), indeed, it benefits from them, but is not restricted to them. The concept can additionally be applied "to city attributes such as *resilience*, *cohesion* or *identity*" and "non-urban communities" (Carrillo, 2015, p. 2). Furthermore, knowledge-based development is not reducible to poles of highly creative people, but includes them. The same holds for ICT technologies: "It may be fortunate to count on, but it can eventually do without, state-of-the art digital grids" (Carrillo, 2015, p. 3). Knowledge-based development is implemented to become globally competitive and to establish sustainability (Yigitcanlar, 2009).

3.2 Indicators of the Country's Development Status

To analyze a country's development status on its way to become a knowledge society, three generally accepted indicators come into question (Linde & Stock, 2011):

- Human Development Index (HDI),
- ICT Development Index (IDI) and
- Networked Readiness Index (NRI).

The HDI is an index published annually by the Human Development Report Office (HDRO) of the United Nations Development Program (UNDP). It is used to present the human development of a country by three dimensions: a long and healthy life (measured by life expectancy at birth), access to knowledge (measured by mean and expected years of schooling) and a decent standard of living (measured by gross national income per capita). For each of the analyzed countries, values between 0 (minimum: worst) and 1 (maximum: best) are calculated. Based on these values, the countries fall into one of the categories low, medium, high or very high human development. With that it is possible to compare the country's human developments of altogether 188 countries with each other (UNDP, 2015).

The ICT Data and Statistics Division within the Telecommunication Development Bureau of the International Telecommunication Union prepares a development indicator on country level as well—the IDI which is published annually in the Measuring the Information Society Report. Altogether eleven indicators are evaluated and subsequently grouped into three categories: access (to ICT), use (of ICT) and skills (to handle ICT). Access data include fixed-telephone and mobile-cellular subscribers, the international internet bandwidth as well as the percentage of households with a computer and with internet access. ICT use is in contrast among others measured by the percentage of individuals using the internet, fixed-broadband and active mobile-broadband subscriptions. The third part—the ICT skills—are among others calculated using the adult literacy rate and the ratio of gross enrolment. The IDI scale reaches from 0 to 10, where 0 is the worst and 10 the best, whereby the indexes are weighted (access and use 40 percent each, skills 20 percent). Using these values, the countries are classified into one of the categories: low, medium, upper or high ICT development. In 2015, the IDI was calculated for 167 countries (ITU, 2015).

Since 2002, the World Economic Forum publishes in cooperation with INSEAD the Global Information Technology Report and with that the NRI. This indicator consists of four sub-indexes: the ICT-environment, the readiness of a society to use ICT, the actual ICT-usage and the impact of ICT. Each of them consists of two or three pillars in turn—that comes to a total of ten pillars, like infrastructure, individual, business or government usage or economic and social aspects. These pillars contain overall 53 individual indicators where the most data source from international organizations, like the International Telecommunication Union (ITU), the United Nations Educational, Scientific and Cultural Organization (UNESCO), other UN agencies and the World Bank. The NRI scale reaches from 1 (worst) to 7 (best) and analyzed 139 countries in the year 2016 (Baller, Dutta, & Lanvin, 2016).

Looking at the values of the HDI 2014 analysis, most of the listed Arab Gulf countries are part of the category very high human development. Only Oman falls into the category where human development is characterized just as high. Qatar reaches with a value of 0.85 rank 32 in comparison to 187 other countries. When

Table 3.1 HDI (2014), IDI (2015) and NRI (2016) values for the GCC states

Country	HDI 2014	Rank	IDI 2015	Rank	NRI 2016	Rank	Rank Arab Gulf region
QATAR	0.85	32	7.44	31	5.20	27	1
UAE	0.84	41	7.32	32	5.30	26	2
Bahrain	0.82	45	7.63	27	5.10	28	3
Saudi Arabia	0.84	39	7.05	33	4.80	33	4
Kuwait	0.82	48	6.83	46	4.20	61	5
Oman	0.79	52	6.33	54	4.30	52	6
Arab region	0.69		5.10				
World	0.71		5.03				

Data sources Baller et al. (2016), ITU (2015), UNDP (2015)

comparing the Arab countries, this is the best result for the region. The other states are almost ten ranks behind. The mean value of the Arab region (0.69) and the world (0.71) are also both below Qatar's value (Table 3.1). Compared to previous years, Qatar improved its human development continually.

The IDI 2015 results differ a little bit from that. Here Qatar reaches with a value of 7.44 the 31st place in the overall rank so that the country is second-placed in the Arab comparison. The country does best in the access and the use category. Just Bahrain reaches a better overall result with a value of 7.63 and the 27th rank. Qatar, Bahrain and also the UAE are concerning ICT on the level of high developed countries. All values of the listed Arab countries are significantly above the Arab and world mean. In this index, Qatar has improved during the last years as well.

In a regional comparison the UAE achieves the best NRI 2016 results (rank 26). Qatar is just one place behind on rank 27 with a value of 5.20. Most of the points are scored in the environment, usage and impact subindex. It is not possible to give any comparison on Arab or world level because no data is available.

An overall evaluation of all indexes and all Arab countries leads to the result that Qatar performs best (with one first and two second places). The average rank—calculated by summarizing the HDI, IDI and NRI rank of the respective country and a following division by 3—is 30 while that of the UAE is at 32, yet and those of the others are even lower. These result shows that the technological basis for a knowledge society exists and that there are also some visible effects these days.

Between the chosen indicators, there are some correlations for all countries on a global scale: According to Pearson's correlation, the development of a country (represented as HDI) and of its ICT development (represented as IDI) interact positively (Linde & Stock, 2011). HDI and NRI act similar so that a high development of a country itself (following HDI) is indicative of a high development of a country as an information society (following NRI) and vice versa (Peña-López, 2006). Likewise, between the information society indicators IDI and NRI there is also a strong correlation of 0.919 (Pearson). Which indicator affects the other one

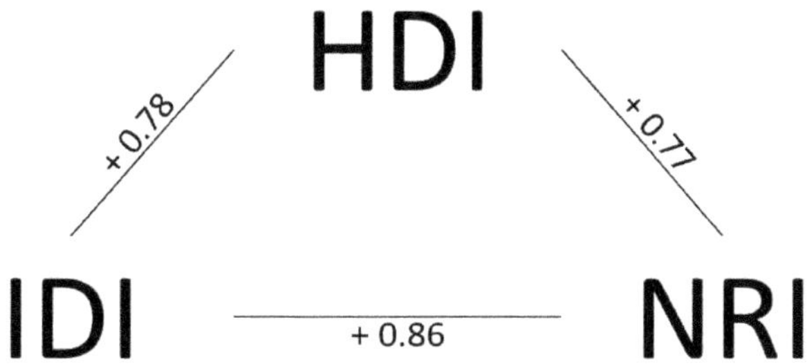

Fig. 3.1 Pearson correlation of HDI, IDI and NRI regarding the GCC states

cannot be determined, but we can make the conjecture that the country's development and its development towards an information society influence each other (Linde & Stock, 2011).

Concerning the GCC countries there is a high correlation between HDI and IDI (Pearson: +0.78) which means that the higher the development of a GCC country, the higher the development of the country's information and communication technology (Fig. 3.1).

Between the human development status (HDI) and the development of the country as an information society (NRI) there is a slightly lower correlation (Pearson: +0.77). So a high human development in the GCC states does translate to an informationally highly developed society. The correlation value of the two information society indexes (IDI and NRI) is as expected still high: +0.86. Hence, we can clearly say that the IDI and the NRI correlate with each other concerning the values of the regions' countries.

After we outlined the main information about Qatar, its characteristics as well as its development status, we will expand this database by collecting data. For this, we will use different methods that were applied both in the field (rapid ethnographic field study and interviews) and on the internet (desktop research and e-Government study) and that shall end up in a theory generated by the method of the grounded theory and with the help of a SWOT analysis. How we proceed in detail, will be described in the following chapter.

References

Baller, S., Dutta, S., & Lanvin, B. (2016). *The global information technology report 2016: Innovating in the digital economy*. Geneva, CH: World Economic Forum.

Bell, D. (1999). *The coming of post-industrial society: A venture in social forecasting*. New York, NY: Basic Books.

Carrillo, F. J. (2015). Knowledge-based development as a new economic culture. *Journal of Open Innovation: Technology, Market, and Complexity, 1*(1), 1–15.

Carrillo, F. J., Yigitcanlar, T., García, B. C., & Lönnqvist, A. (2014). *Knowledge and the city: Concepts, applications and trends of knowledge-based urban development*. New York, NY: Routledge.

Castells, M. (1989). *The informational city: Information technology, economic restructuring, and the urban-regional process*. Oxford, UK: Basil Blackwell.

Castells, M. (2010). *The information age: Economy, society and culture* (2nd ed.). West-Sussex, UK: Wiley-Blackwell.

El-Haddad, Y. (2003). *Major trends affecting families in the Gulf countries*. Sakhir, BH: University of Bahrain.

Ergazakis, K., Metaxiotis, K., & Psarras, J. (2006). An emerging pattern of successful knowledge cities' main features. In F. J. Carrillo (Ed.), *Knowledge cities: Approaches, experiences, and perspectives* (pp. 3–15). New York, NY: Butterworth Heinemann.

Fietkiewicz, K. J., & Stock, W. G. (2015). How "smart" are Japanese cities? An empirical investigation of infrastructures and governmental programs in Tokyo, Yokohama, Osaka and Kyoto. In *Proceedings of the 48th Hawaii International Conference on System Sciences, Jan 5–8, 2015, Kauai, Hawaii* (pp. 2345–2354). Washington, DC: Science, IEEE Computer.

Florida, R. (2002). The rise of the creative class: Why cities without gays and rock bands are losing the economic development race. *The Washington Monthly, 34*(5), 15–25.

Giffinger, R., Fertner, C., Kramar, H., Kalasek, R., Pichler-Milanović, N., & Meijers, E. (2007). *Smart cities: Ranking of European medium-sized cities*. Vienna, AT: Centre of Regional Science.

Hall, R. E., Bowerman, B., Braverman, J., Taylor, J., Todosow, H., & von Wimmersperg, U. (2000). The vision of a smart city. In *Proceedings of the 2nd International Life Extension Technology Workshop*. Paris, FR: 2nd International Life Extension Technology Workshop.

ITU. (2015). *Measuring the information society report 2015*. Geneva, CH: International Telecommunication Union.

Knight, R. V. (1995). Knowledge-based development: Policy and planning implications for cities. *Urban Studies, 32*(2), 225–260.

Linde, F., & Stock, W. G. (2011). *Information markets: A strategic guideline for the I-Commerce*. Berlin, Germany, New York, NY: De Gruyter.

Masuda, Y. (1980). *The information society as post-industrial society*. Tokio, JP: Institute for the Information Society.

Peña-López, I. (2006). Networked readiness index vs. human development index. *ICTlogy, 30*.

UNDP. (2015). *Human development report 2015*. New York, NY: United Nations Development Programme.

Webster, F. (2006). *Theories of the information society*. Abingdon, UK: Routledge.

Yigitcanlar, T. (2009). Planning for knowledge-based urban development: Global perspectives. *Journal of Knowledge Management, 5*, 228–242.

Yigitcanlar, T., & Velibeyoglu, K. (2008). Knowledge-based urban development: The local economic development path of Brisbane, Australia. *Local Economy, 23*(3), 195–207.

Chapter 4
Methods

Abstract How does one investigate the knowledge-based development of a city or a region? For this study, various methodological approaches were used to obtain the needed data as well as to arrange and to analyze the findings. The rapid ethnographic field study approach enables the researchers to witness first-hand the investigated region, the on-site experts' interviews enrich this experience with a view on the city's potentials and weaknesses from a different perspective. The outcomes of these qualitative approaches are completed with data obtained through desktop research, including bibliometrics and social media metrics, and, finally, evaluated with help of the *SWOT*-analysis (an acronym for *Strengths, Weaknesses, Opportunities, and Threats*). This mixture of qualitative as well as quantitative data and the recursive theory development constitutes the grounded theory method, which was the base for the applied research framework.

Keywords Grounded theory · Rapid ethnography · Field study
Qualitative interviews · Social media metrics · Informetrics · SWOT analysis

4.1 Grounded Theory Method

The grounded theory method, introduced by Glaser and Strauss (2009 [1967]), is a method to inductively develop a theory during research with continuous regard to the retrieved data. By data we mean all kinds of materials like interviews, observations, videos and other documents (Corbin & Strauss, 2015). A grounded theory is established by "moving from the particular to the general" (Mason, 2002, p. 180). Rather than to verify old theories, entirely new theories are generated step by step from the collected data relating to the research question (Mansourian, 2006). This approach corresponds with the aim of the current study, which is to generate theories on Qatar's knowledge-based development from data collected on-site or retrieved during desktop research. Furthermore, combining research approaches based on qualitative and quantitative data is not new to the domain of human geography, a research area where the knowledge-based development of cities and

J. Gremm et al., *Transitioning Towards a Knowledge Society*,
https://doi.org/10.1007/978-3-319-71195-9_4

regions can be located. "Contemporary human geographers study places, people, bodies, discourses, silenced voices and fragmented landscapes. The research questions of today's human geographers require a multiplicity of conceptual approaches and methods of enquiry. Increasingly, the research methods used are qualitative ones intended to elucidate human environments, individual experiences and social processes" (Hay, 2010, p. 2).

A grounded theory is generated by the process of data collection called theoretical sampling, which combines the operations "collection," "coding" and "analysis" of data. All of these operations should be executed jointly and repeatedly (Glaser & Strauss, 2009 [1967]). "Here, analysis is necessary from the start because it is used to direct the next interview and observations" (Corbin & Strauss, 1990, p. 6). This means that the research questions can be improved step by step in the course of the analysis. The decision which data will be collected is not fixed from the beginning. During the process of theory construction, all data needed to be collected additionally is identified. For example, all unexpected problematic issues or discussion aspects arising during one expert interview, can be permanently integrated in the interview guidelines and addressed in all following conversations.

Beside the data collection and analysis, the researcher is permanently required to conduct coding, due to the steady arrival of new data. For the process of coding the comparative analysis is utilized. The aim is to construct the grounded theory (Glaser & Strauss, 2009 [1967]). The available data—old and new—which are treated as concepts, are constantly compared, interpreted and grouped or regrouped to form categories and their properties. This serves as a basis to formulate hypotheses (Corbin & Strauss, 1990). Categories, properties and hypotheses are elements of a theory. Categories are elements of the theory itself, properties are elements of a category, and hypotheses are relations among categories and their properties. Those hypotheses need to be justified during the theory construction in order to generate a valid model with regard to the research question (Glaser & Strauss, 2009 [1967]).

The grounded theory method is suitable for research on urban knowledge-based development, since it is not restricted to a certain discipline, but rather an appropriate method for all fields (Glaser, 1992). In addition to the collection and coding of qualitative data, retrieved was quantitative data in form of official statistics and informetric analyses on literature search services (Web of Science and Scopus) as well as social media (Facebook and Instagram). The variety of primary and secondary data constitutes a strong base for the theory development.

4.2 Rapid Ethnographic Field Study

The collection of qualitative data was possible due to the ethnographic field study. Ethnography is a "research method dependent on direct field observation in which the researcher is involved closely with a social group or neighborhood" (Hay, 2010, p. 187). A typical ethnographic field study aims at data collection and studying people, their social meaning as well as their everyday life in their natural environment without influencing the existing surrounding (Brewer, 2000). In general, it is important to observe the object of study for an extended period of time, which helps to understand the investigated nations or cultures (Malinowski, 1992). For the purpose of this investigation, the rapid form of ethnographic field study was applied, which is a modification of the traditional approach. This kind of rapid ethnography pursues the same goal and uses a similar combination of different field methods, however, it takes place in a limited time. Millen (2000, p. 280) summarizes his key ideas for this approach:

> First, narrow the focus of the field research appropriately before entering the field. Zoom in on the important activities. Use key informants such as community guides or liminal group members. Second, use multiple interactive observation techniques to increase the likelihood of discovering exceptional and useful user behavior. Third, use collaborative and computerized iterative data analyze methods.

The method is described as naturally time-deepening, which makes it possible to accomplish different activities simultaneously in a fast way so that the researchers can study more in less time. In contrast, traditional wide-angle focus approaches for ethnographic studies take a lot of time to reveal useful results. They boost the costs, potentially distract from the research focus and are often not helpful, when research time is limited or the studied technology is rapidly changing. The narrower view through a telephoto lens, like in the rapid ethnographic field study, leads to more useful applicable data in a shorter time. Basis for a successful study is that the research question is well defined and the camera is well positioned (Millen, 2000), so that all the important but no redundant information is gathered.

This method is suitable for the current study, since an extensive groundwork was already laid in previous investigation on the same research topic and region (Kosior, Barth, Gremm, Mainka, & Stock, 2015). In order to get an even better understanding of the state of Qatar, the people living there, their culture and daily life, this groundwork was completed with research on the latest region-specific literature. All places to visit were well researched prior to the field study. During the nine days on-site, pictures or videos of places strategic to the knowledge-based development were taken. Also, people living in Qatar were observed and their behavior and social environment was recorded by the team of three researchers.

Furthermore, all potential interview partners were contacted far in advance. This way it was possible to conduct all interviews on-site during a relatively short period of time. The potential interview partners were contacted via email. They had to fulfil certain pre-defined selection criteria. This method is the so-called criterion sampling, which means "choosing all cases that satisfy some predetermined standard" (Hay, 2010, p. 186). The interview partners had to be knowledge workers (working in a field of the knowledge economy) and natives as well as expats living in Qatar. The interviews were semi-standardized, meaning that there was an interview guideline with pre-defined questions; however, with no strict order and with the possibility for discussion of further, unplanned topics. The pre-defined questions were based on an extensive research, elaborated in the following subchapter.

4.3 Qualitative Interviews

The "qualitative research is concerned with elucidating human environments and human experiences within a variety of conceptual frameworks" (Hay, 2010, p. 3). For the purpose of this study, the focus of the qualitative research lies on environments and experiences linked to the knowledge society and knowledge-based development of Qatar. The primary qualitative data was gathered during semi-structured interviews. This type of interview has "some predetermined order, but which nonetheless has flexibility with regard to the position/timing of questions. Some questions, particularly sensitive or complex ones, may have standard wording for each informant" (Hay, 2010, p. 196).

The interviews were conducted during the rapid field study in Doha from February 16, 2016 until February 24, 2016. In total 13 experts were interviewed on the development of knowledge society in Qatar. The interview guidelines included 65 questions, which covered the following topics:

- diversification of knowledge-based industries,
- creativity and knowledge society,
- e-Government,
- knowledge-based development,
- universities and other institutions of higher education,
- transition of graduates into the labor market,
- libraries, science parks and research funding, and
- SWOT (strengths, weaknesses, opportunities and threats regarding Qatar's way into the knowledge society).

Each topic and the corresponding questions will be elaborated in more detail in the following.

4.3.1 Diversification of Knowledge-Based Industries

Why is the diversification of knowledge-based industries relevant for this investigation? "As the countries of the Gulf Cooperation Council (GCC) strive to compete in the 21st century, they face a number of challenges, including fluctuating oil and gas prices, economic diversification and the need to build up human capital and create jobs for GCC nationals" (Tadros, 2015, p. 1). Especially the decline in oil and gas prices as well as the idea of diversifying the economy brings up one of the first interview questions: *What is the state of the art and what are the development plans of the oil and gas industry?*

There are certain types of companies having their headquarters in the typical cities of the knowledge society. One of them is "a mixture of service providers and industrial companies, the information economy, to which belong industries such as computer manufacturing, software development, telecommunication, and Internet firms (e.g., search engine providers) as well as information service companies (producers and providers of information services)" (Stock, 2011, p. 977). Based on this notion, the question about Qatar's industry diversification towards the ICT economy arose: *What is the state of the art and what are the development plans of computer science and information science, the respective industries (e.g., information and communication technology) and their application in companies as well as administrations (e.g., knowledge management)?*

An important aspect of the modern urban development became the sustainability and eco-friendliness. "Sustainability is an established goal of future urban developments" (Monfaredzadeh & Krueger, 2015, p. 1112). It became necessary "to create a governance mechanisms and the institutions to move the world toward a green economy" (Puppim de Oliveira et al., 2013, p. 138). Therefore, it is to expect that the renewable energies industry will be booming in the upcoming years. It is another possibility to diversify the country's economy. This notion lead us to the next question, *what is the state of the art and what are the development plans of the renewable energies industry (especially, solar-sourced power)?*

Another type of companies typical for cities of the knowledge society are creative or cultural services. Some of them "provide—in the context of soft location factors—cultural facilities (e.g. theater, opera, etc.)" (Stock, 2011, p. 977). Hall (2005) speaks of cultural/creative, business and tourism services that are highly synergistic with each other. For example, hotels and conference or exhibition centers fit into the business and tourist category, museums and galleries into the creative/cultural and tourism sector, whereas advertising is a creative and business category (Hall, 2005). MICE (meetings, incentives, conferences, events) industry promotes (business) tourism as well as (international) information exchange (Fietkiewicz & Stock, 2014, p. 48). This brings up the next two questions: *What is the state of the art and what are the development plans of the MICE (meetings, incentives, conferences, events) industry?* And, *what is the state of the art and what are the development plans of the tourism industry?* The inquiry about the industry diversification was completed with a general question, whether there are *any other (scientific and technological) disciplines and industries to be build up?*

4.3.2 Creativity and Knowledge Society

Creativity constitutes an important component of urban knowledge-based development. According to Florida (2002), creativity, or rather the so-called "creative class" fosters the economic growth of a region. Therefore, several questions concerned this aspect. In the interviews, this topic was introduced with the question—*According to Richard Florida, the "creatives" (artists, scientists, etc.) are the driving force of present and future cities. How do you estimate the situation of creatives in Qatar?*

Florida (2002) believes that those belonging to the creative class are well educated and responsible for corporate profits and economic growth. However, to attract more human capital including talents and creative people like, e.g., scientists, it is important for a country or city to be open and diverse (Florida, 2005). This assumption was reflected in the questions—*Is there an innovative "atmosphere" in Qatar? Is there an "atmosphere" of openness and tolerance in Qatar?*

Being part of the creative class scientists bring economic benefits to a region. Those benefits can be measured with the help of the indicator *research*. It includes the input and output indicators, which reflect diverse research activities (Godin, 2007). In this study, scientists were defined as the *research input* indicator, when stating the question—*What are the roles of scientists in Qatar?* The follow-up question—*Do they only teach, or do they perform research as well?*—arose due to outcomes from the previous studies. These indicated that private universities in the Gulf region sometimes only revolve around teaching, while almost entirely neglecting research. Furthermore, in modern knowledge societies the scientists often give advice to political decision-makers (Maasen & Weingart, 2005), therefore, the question arose—*Do they give advice to politics and administration?*

Since knowledge societies focus not only on scientific, but also on technological and medical knowledge (Böhme, 1997), similar questions were formulated with regard to the technology and medicine: *What are the roles of engineers in Qatar? Do they teach only, or do they perform research as well? Are they concerned in construction processes in Qatar? Do they give advice to politics and administration? What are the roles of physicians in Qatar? Is there any high-class medical education? What are the state of the art and the future of hospitals?*

For the *research output* indicator, knowledge and inventions (Godin, 2007) as well as social media metrics were chosen as indicators of scientific impact (Haustein, Sugimoto, & Larivière, 2015). The corresponding questions were: *How visible are Qatar's research results in terms of publication and citation counts, but also regarding the altmetrics (tweets, blog posts, mentions of newspapers)? Are there any core themes? Is there any specialization of Qatar's research?* And, *how visible are Qatar's engineering results in terms of patents? Are there any core areas? Is there any specialization of Qatar's engineering?* Those questions concerning core themes and specializations resulted from the fact that "a region is dependent on only few topics and few institutions. Learning about these topics and institutions can be very important to regional policy" (Altvater-Mackensen et al., 2005, p. 526).

Just as important as the production of knowledge is its distribution (Kavanaugh et al., 2012, p. 480):

> Citizens are increasingly relying on social media for communication with their friends, work teams, social organizations, and government. The capabilities to facilitate interpersonal and group interaction provide new and unique opportunities for community leaders, elected officials, and government service providers to inform, and be informed by, the citizenry. Twitter, Facebook, Flickr, YouTube, and other services with user-generated content have made a staggering amount of information available.

This raised the last group of questions regarding the creativity and knowledge society: *How is the exchange of everyday knowledge organized? Do people in*

Qatar use social media (e.g., Facebook, Twitter, etc.), messaging services (e.g., WhatsApp), classical media (telephone, SMS), or do they prefer to communicate face to face? Is there an institution like dewaniya in Qatar? Dewaniya is a traditional meeting place in Kuwait. In a house or a tent, people meet to cultivate contacts (Gremm, Barth, & Stock, 2015). We were interested if there exist similar gathering places (or majlis) in Qatar, where face-to-face communication is the only form of knowledge exchange.

4.3.3 E-Government

The next group of questions concerned the state of Qatar's e-Government. E-Government and e-Governance are essential parts of urban development in the knowledge society context (Stock, 2011). Concerning the e-Governance, hence, the political willingness of the authorities to build up a knowledge society, Qatar's e-Government 2020 Strategy lays its focus on the people and promises that everyone will benefit from it (ictQatar, 2014). In order to ascertain whether this attempt has been successfully implemented until now and whether the citizens are included in the process of improving the country, we wanted to know: *Is there in Qatar a fruitful knowledge exchange between citizens and administrations?*

Since the 1960s, the involvement of scientists in the political process has proceeded. They are called in as experts to support political positions and decisions (Maasen & Weingart, 2005). Consequently, we wanted to know: *Do the political and administrative elites consult scientists or engineers while discussing (master) plans and other decisions?* If they do not, the establishment of new plans might be restricted by limited views. Another possible restriction is the country's culture. As written in the Holy Qur'an, Allah has full knowledge. Since the government and its people in the Gulf region lead their lives fully devoted to their culture and religion (Wiseman & Alromi, 2003), the following question arises: *Do the Arabian culture and the Moslem religion support knowledge-based development?* Furthermore, the development of e-Government, the maturity of governmental website, was evaluated by the researchers based on objective criteria catalogue (see Sect. 4.5), and not inquired during the interviews.

4.3.4 Knowledge-Based Development

It is not a secret that the country is already aiming at a change. Qatar's Gross Domestic Expenditure on Research and Development (GERD) per GDP is 2.8 percent (Kosior et al., 2015). Now, we wanted to learn the opinion of our interviewees: *Do the political leaders adequately fund the emerging knowledge society?*

According to Stock (2011), a knowledge-based society has to attract four types of companies: capital-intensive service-providers, knowledge-intensive high-tech industries, service providers as well as creative enterprises. We are interested whether Qatar is able to attract these companies and stated the question: *Are there or will there be in future knowledge-based companies (e.g., high-tech industries, consultants, financial services) with workplaces for knowledge workers?* If workplaces are available, it is important that these positions can be filled with properly educated knowledge workers. For this purpose, the country needs to provide the best education to its citizens. As we learned that students in the GCC states are still lagging behind in terms of knowledge, skills and competencies (Hoteit, Tuerpitz, & Luelsdorf, 2015), we formulated the next question: *Are there or will there be in future enough well-trained teachers in elementary and high schools?* But not only the availability of excellent teachers needs to be considered. The family is said to be a significant factor for the Qataris' learning motivation (Hvidt, 2015)—so, if the parents are mostly not very well educated or do not encourage their children to reach for the stars, will their children aim at success? From this vantage point we asked: *Is it a goal for Qataris and for expats to see their children well-educated?*

4.3.5 Universities and Other Institutions of Higher Education

Universities and other institutions of higher education are the building stones of a knowledge society, since they qualify the future knowledge workers. The Education City located in Doha incorporates many educational facilities, including branch campuses of universities worldwide. To get a complete picture of the concept of branch campuses in Education City we asked *who the owner of the branch campuses is*. The follow-up question—*Is it possible to earn a doctorate (Ph.D.) degree?*—was posed to find out about the range of degrees that can be achieved by students in Qatar. While we know that especially the private universities and higher education institutions in the Gulf region mostly concentrate on teaching rather than performing research, we do not know about the standard of research and teaching. Thus, the questions—*Is the level of research at Qatar University or at a branch campus in Education City comparable to American, British, Australian or European standards?* And, *Is the level of education at Qatar University or at a branch campus in Education City comparable to American, British, Australian or European standards?*—immediately suggested themselves. We also wanted to know, whether there are any differences between Qatar University, as the public university of the country, and the higher education institutions in Education City, which are all private universities, and asked our interviewees—*Are there any differences between Qatar University and the branch campuses in Education City in terms of resources, level of research, and level of education? Are universities adequately funded? What is Qatar's GERD per GDP?* Here, we directly asked our interviewees whether they are aware of the amount that Qatar spends on R&D and whether universities benefit from these investments. Additionally, we wanted to know about the situation within the higher education institution itself—*Are the organizational and financial resources (e.g., for performing research, for trips to conferences, etc.) adequate?*

"Countries of the Arabian Gulf, in particular, have emphasized free zone development over the past decade" (Rao, 2000, p. 246). Our question—*Do universities and other higher education institutions pay taxes?*—refers to this topic. *Do the conditions in the free zones also hold for the foreign universities?* Since there are many expats in Qatar, we wanted to know what actually attracts them to this region. We approached this topic by asking—*Do staff members pay taxes? Is the salary of staff members attractive?*

The Gulf region has the highest concentration of branch campuses worldwide (Becker, 2010). This striking discovery gave rise to the question—*Are there any problems to organize branch campuses? What are their advantages and disadvantages? According to THE TIMES, Qatar University is the most international university in the world (in terms of international staff, students and cooperation). What are the advantages and disadvantages of such internationalization?* This question was caused by Times Higher Education, which revealed a ranking of the world's most international universities in 2016 with Qatar University on the top of the list, hence, having the best international outlook (Bothwell, 2016).

A tolerant society is open "to all ethnicities, races, and walks of life" (Florida, 2003, p. 10). This leads to the question: *What are the advantages and disadvantages of multi-national research teams?* The Westernization or Americanization of the Gulf tertiary system can be linked to the branch campuses operating in Education City (Witte, 2010). Since the literature indicates that this might be a problem, we asked—*Due to internationalization, is it there any loss of identity for Qatari? Is there a danger of "Americanization" or "Westernization" of education?* The number of expatriate faculties exceeds the number of Qatari faculties in public and, especially, in private higher education institutions (Kosior et al., 2015). To learn the reasons for this imbalance we posed the questions—*How many staff members of Qatar University and of the branch campuses of Education City are Qataris and how many are expats? Why are only few Qataris working at institutions of higher education?*

When thinking about the dependence of Qatar on foreign workers, an interesting question is—*How many Qatari and how many expat children study in Qatar?* The answer might reveal whether there will be enough Qatari students in future that will enter the local labor market after graduation. And this is again an important hint to the question, whether Qataris or expats will be employed in leadership positions. Since foreign workers in Qatar depend on the country's *kafāla* system that might bear some disadvantages to the children of the foreigners due to restricted employment, we wanted to know—*Do expat children face problems because of the kafāla system?* The *kafāla* system was used to control the migrant laborers and was heavily criticized by human rights organizations. However, in December 2016 this system was modified by Qatar's government.

Just like there are differences between Qataris and expats, there are also differences between males and females. In the Gulf region the number of female students and graduates is higher than the one of male students and graduates (Kosior et al., 2015). In our interviews we wanted to find an explanation for this situation—*There is an imbalance between male and female students and graduates (in favor of women) in Qatar. What are the reasons? Is this fact problematic for the universities?* According to Gill (2008), Qatari men can travel overseas to graduate. This fact raised the next questions—*Do some Qataris avoid to study in Qatar? Why do they study away from home?* One of the reasons why Qatari girls go to universities or other higher education institutions is that they would have to stay home otherwise (Kosior, Barth, Gremm, Mainka, & Stock, 2015). Since we wanted to get more information on the reasons why females and males decide to study, we asked—*What are the motives of female students to study? And the motives of male students to study?*

Furthermore, we wanted to discuss the differences between the public and the private universities in Qatar with our interview partners. Qatari nationals do not have to pay for higher education, when they are good enough to enroll at the national universities (Moini, Bikson, Neu, & DeSisto, 2009). What about the expat children? Are they provided with free education? And what about private universities? With all these aspects in mind, we asked our interviewees—*Are there tuition fees for Qatari and expat students?* Vora (2014) drew our attention to the phenomenon that Qatari students study in most cases funded by different companies or Qatar Foundation acting as a sponsor. Owing to this support, the students commit themselves to show the sponsors their gratitude by working for them for a period of time. Suggested by this idea, we tried to find out what this means and wanted to compare the already acquired results with our findings, so we asked—*Are most of the Qatari students funded by sponsorships? Who are the sponsors? What is the students' (graduates') service in return?*

We know that gender-segregated education exists in Qatar. Qatar University, for example, offers separate courses for female and male bachelor students (Kosior et al., 2015). Since we only know that this segregation exists at Qatar's public university, but do not have any information about the private universities in Education City, we posed the questions—*Are there gender-separated studies in Qatar? Why? What are the advantages or disadvantages of gender-specific studies?*—to get a complete picture of this subject.

Ulmer (2009) mentions problems among Qatari students with the English language, which is the language of instruction of the branch campuses in Education City. To know if this is a general issue in Qatar's higher education institutions we brought up the question—*Are there any problems with English as teaching language?*

4.3.6 Transition of Graduates into the Labor Market

Since in Qatar the average year 9 student is 3–5 years behind the students in other OECD or more educationally advanced cities or countries (QF, 2013), we wanted to know how is their status after finishing the university and asked—*Are the graduates well-educated when they leave the higher education institutions?*

The work-life balance for women in the UAE seems to be perfect because working hours are between seven and eight hours on average, whereby the commute and the duties they have to fulfill at home are already included (Forster, Al Ali Ebrahim, & Alma Ibrahim, 2013). Since the conditions are so good, we wanted to know—*Are there any problems for the graduates to search for and to find an employment in the public sector?* Furthermore, to learn more details about the conditions of Qatar's working environment, we additionally asked—*How are the working conditions (work-life balance, payment) in the public sector?* With regard to our research topic concerning the knowledge society we also wanted to know—*Are there knowledge-intensive jobs in the public sector?* In order to draw a comparison concerning the job opportunities, the working conditions and the availability of knowledge-intensive jobs, we asked the same questions for the private sector. *Are there any problems for the graduates to search for and to find a job in the private sector? How are the working conditions (work-life balance, payment) in the private sector? Are there any knowledge-intensive jobs in the private sector?*

According to Al-Kitbi (2010), there is progress in gender equality in education, but women still have some difficulties getting into leadership positions. This is why we came up with the following questions—*Are there any gender-specific differences to search for and to find employment? Is it possible for all graduates (regardless of gender) to reach leadership positions?* Beside the difference between the genders, we also wanted to know–*Are there any differences for Qatari graduates and expat graduates to search for and to find employment?* Since in Qatar "national male as well as female graduates prefer jobs in public services" (Kosior et al., 2015, p. 36), we tried to ascertain the reasons. We asked–*What are the motives of male and female graduates to apply for a job in the public (or private) sector?*—to find out the advantages and disadvantages of both sectors. More than 25 percent of the under 25 year olds in the GCC are unemployed. This is caused, among other reasons, by lacking knowledge, skills and competencies. Work experience programs at schools and universities like internships might help to solve this problem (Hoteit et al., 2015). This is why we wanted to know—*Do institutions of higher education cooperate with companies in Qatar (e.g., for internships or Bachelor theses)?* We know about the Nationalization programs in the Gulf region, where expats shall be replaced by locals—in our case Qataris—to localize the countries' labor market (Godwin, 2006). *Is there any success of Qatarization?*—was the question we were interested in concerning this problematic aspect.

4.3.7 Libraries, Science Parks and Research Funding

Librarianship in Qatar is placed at the "forefront of the country's future plan" (Johnston & Williams, 2015, p. 86). Thinking about this new position of librarians in Qatar we wanted to know *What roles do librarians and other information*

Fig. 4.1 The Library of Qatar University

professionals play in the development of Qatar's knowledge society? And concluding that major expertise would be needed for this, we asked—*Are librarians in Qatar well-trained?* From previous studies we already know the National Library in Qatar, and the Library of Qatar University (Fig. 4.1). However, we do not know anything about the library's tasks, therefor we asked—*What is the function of the National Library?* Also, one of our previous studies showed that libraries are not frequently attended by the citizens (Kosior et al., 2015, p. 24). Therefore, we asked —*Do people (students, but also citizens) really accept and use libraries?*—to get a detailed understanding of the behavior of people living in Qatar. Additionally, we asked—*Are there any differences between Qataris and expats?* Since we know about the existence of information literacy instructions given by librarians in other countries, like for example Canada (Henkel, 2015) and the U.S. (Henkel & Stock, 2016), we wanted to know—*What is the state of information literacy among Qatar's population?*—And, in addition, *is there any information literacy instruction in libraries?* In order to find out about the citizens' interest and acceptance we asked—*Are such courses well-visited and popular?*

Beside libraries, science parks play a significant role in modern knowledge societies. They attract companies to settle down, ensure interaction between firms and academic institutions and, hence, forge a "science—industry relationship" (Hansson, Husted, & Vestergaard, 2005, p. 1039). Since the Arab region has a rather low number of such science parks, but at the same time some of the big ones are located in Qatar (Qatar Science and Technology Park and additionally Sidra Medical and Research Center), we are interested in what exactly they are doing—*What are the tasks of Qatar Science and Technology Park (QSTP) and of Sidra Medical and Research Center (SMCR)?* In addition, we are interested in our interviewees' opinion on the question—*Is there need for further science parks?* In most countries there is a research foundation supporting research and science, like the Deutsche Forschungsgesellschaft (DFG) in Germany, the National Science Foundation (NSF) in the United States and the Qatar National Research Fund (QNRF) in Qatar. For a better understanding of their function, we asked—*what are the tasks of Qatar National Research Fund (QNRF)?* Another question that we were interested in was—*Are there university–government–business ("triple helix") partnerships?*—since, according to Abduljawad (2015, p. 58), "University-Industry-Government (UIG) collaborations are major contributors to knowledge assets at organizational and national levels".

Finally, the economic development in cities of knowledge societies can be fostered by enhanced entrepreneurial activity, in context of knowledge-based development especially by knowledge-intensive start-ups. "A high level of new firm creation significantly contributes to regional economic vitality and is a major signal of a dynamic economy" (Lee, Florida, & Acs, 2004, p. 879). There are various factors influencing the level of entrepreneurial activity, however, we focused only on the question, what is the state of entrepreneurship in Qatar and asked—*Are there knowledge-intensive start-up's (e.g., ICT companied, Internet services) in Qatar? Is there any e-Entrepreneurship?*

4.3.8 Qatar's Way into the Knowledge Society

Finally, we asked our interview partners (as all of them are expats): *Why did you come to Qatar? How long have you lived here? And how long do you intend to stay?*—to learn more about the motivation of the expats in the region. To assess the already achieved goals and the not yet implemented development aspects of Qatar in becoming a knowledge society, we asked—*How do you estimate the development of Qatar's knowledge society in terms of SWOT?*—and resolved the acronym of this common marketing tool by asking: *What are Qatar's strengths, weaknesses, opportunities and threats?* (Ferrell & Hartline, 2014). On this basis and on all other answers to our questions, whether they were received from our interviewees or other applied methods in our study, we will weigh Qatar's chances of already being or of becoming a knowledge society.

4.4 Quantitative Data Analysis

Besides the qualitative data that the Grounded Theory Method focuses on, we ascertain quantitative data in form of statistics of Qatar and the number of publications and patents as well as the number of posts on Facebook and Instagram originating in the country. This data helps to reinforce the qualitative information and to make reasoned statements about the topics of interest. The required quantitative data was gathered during online research.

Statistical data ensures comparison, for example, of different institutions. We try to focus on official statistics from ministries whenever possible, because we can assume correctness and integrity of the information. However, working with statistics from the Gulf region is problematic, because the region does not have a system that uses standardized methods for the gathering and reviewing of data. Therefore we had to access different sources to collect the needed statistics. Thereby we met problems like differing ways to display facts and figures, different enquiry periods and missing information.

For quantitative data analysis, we draw on the field of informetrics, which "is the study of quantitative aspects of information in any form, not just records or bibliographies, and in any social group, not just scientists" (Tague-Sutcliffe, 1992, p. 1). Egghe (2005, p. 1311) uses the term "informetrics" as the broad term comprising all -metrics studies related to information science, including bibliometrics (bibliographies, libraries, etc.), scientometrics (science policy, citation analysis, research evaluation, etc.), webometrics (metrics of the web, the Internet or other social networks). While traditionally quantitative investigations of scholarly activity mainly considered publications and citations, it is also possible to include other scientific outputs in a study (Cronin, 2001). In our case the number of patent applications is of great interest. In our study, we focus on descriptive informetrics, which, according to Stock and Weber (2006), include individual documents, subjects, authors, institutes, scientific fields, regions, countries and other individual items. Following the two authors, we were guided by research questions like "What are the core subjects of the publications of Qatar?" and "How many articles does Qatar publish per year?" and "How many patents does Qatar apply per year?"

Furthermore, we include altmetrics, also referred to as social media metrics (Haustein, Larivière, Thelwall, Amyot, & Peters, 2014), which is a subset of scientometrics as well as of webometrics (Priem, Groth, & Taraborelli, 2012). Altmetrics is a new area of research, which focuses on measuring scientific impact derived from social media environments (Priem, Taraborelli, Groth, & Neylon, 2011). Here evaluated are, beside the number of citations, other categories like the number of views, saves, comments and recommendations (Fenner, 2013). Just as we are interested in counting the number of publications and applied patents of Qatar in the traditional way, we also identify the number of posts published by Qatar's ministries and agencies. To measure their impact, we additionally quantify likes, comments and shares and also the number of followers, in order to get an idea about how many people they reach.

The scientific output in terms of publications is an important indicator of our evaluation of Qatar. To receive the number of Qatar's publications we consulted the two renowned scientific databases Web of Science[1] and Scopus.[2] Starting with the publications available in Web of Science, we chose the basic search in the Web of Science Core Collection and restricted our search to publications where the address

[1]Web of Science: http://webofknowledge.com/.

[2]Scopus: http://scopus.com/.

includes either the term "Qatar" or the term "Doha". Web of Science allows to further analyze the results, giving us the opportunity to investigate and rank our findings by publication year and Web of Science category. Since we are also interested in the institutions that publish most of Qatar's articles, we ranked the results by "Organizations Enhanced" to get a first impression and then switched to the advanced search where we used the field "Organizations Enhanced" to analyze the most engaged higher education institutions in greater detail. To receive the number of publications of Qatar included in Scopus, we chose the advanced search. Here, all publications were counted where the affiliation includes at least one of the two terms "Qatar" or "Doha". Publication years and disciplines are provided by Scopus in form of rankings right away, which we used to receive even more information about the publications listed in Scopus.

A second indicator of scientific output is the number of patents of a country. Although there is a patent office for the Arab states of the Gulf, namely the GCC Patent Office,[3] its information and accuracy is limited. To determine the exact number of granted patents filed by Qatar is impossible. Because of that, we analyzed the number of patents of Qatar with the help of the scientific database Scopus. The same advanced search that reveals the number of Qatar's publications registered in Scopus simultaneously displays the number of registered patents of Qatar. Those are not applied in Qatar, but their inventors come from this country.

4.5 E-Government and Government 2.0

In a knowledge society and the so-called informational city, that this society lives in, e-Governance is one important basis for innovation (Yigitcanlar, 2010). E-Governance is a generic term for planning, innovation and funding at city level (Sharma & Palvia, 2010). One of the main goals set in many e-Governance plans is to establish a timely e-Government. Indeed, the increased use of ICT and knowledge management between authorities and citizens or other stakeholders lead to a better and more complex e-Government services, and call on citizens and companies to actively engage in political debate and decision making processes (Gisler, 2001; Kettl, 2002; Sriramesh & Rivera-Sánchez, 2006; Sharma & Palvia, 2010). In our study, we take a deeper look at Qatar's e-Government website *Hukoomi*. We evaluate the portal based on the five pillars model by Fietkiewicz, Mainka, and Stock (2017). Their approach originated from the five-stage model by Moon (2002) and by Hiller and Bélanger (2001). The investigated five pillars of e-Government website include: information, communication, transaction, integration, and participation (Table 4.1). Each one encompasses diverse aspects, which concern services provided by the government for different stakeholders. Each aspect is weighted

[3]GCC Patent Office: http://www.gccpo.org/DefaultEn.aspx.

Table 4.1 Five pillars of e-Government

Pillar I: information	Pillar II: communication	Pillar III: transaction	Pillar IV: integration	Pillar V: participation
Current press releases	Social media services	Online forms	Entry homepage	Online questionnaires
Contact information	Online appointments	Online payments	Cooperation between authorities	Online surveys
Applications	Online feedback	Library services	Intranet	Online community meetings
Push-services	Email-response	Personalized web-portal	Common database	Online voting
Downloadable forms			Common safety measures	
Information on politics, healthcare and services				
English version of the webpage				

Source Fietkiewicz et al. (2017)

appropriate to its importance and complexity. Each pillar can achieve the maximum value of 100 points, hence, to the total maximal value are 500 points.

The first investigated pillar is the information dissemination. It refers to the one-way communication in which information is only transmitted from e-Government to the users. There are two groups of information types we focused on, classified by the information content and the user types the information is for. For the first group, the government websites should offer basic data such as contact information, current affairs, emergency numbers or a city map (Fietkiewicz et al., 2017). Other aspects include, for example, transportation, health care, politics, or push-services. Additionally, the website should be available, despite the country's native language, in English and in the languages of the three most important immigrant groups. For the second group, the websites' informativeness for certain groups of users, e.g. residents, tourists, students, and businessmen should be ascertained.

The second pillar is the (two-way) communication, nowadays strongly linked to social media. This pillar is divided into five parts: (1) social media services, such as Facebook and Twitter, (2) fixing appointments online, (3) email response, (4) official email instead of snail letter mail, and (5) possibility of giving feedback.

The third pillar, transaction, consists of financial and non-financial transactional e-Government services such as renewing driver's license, voter registration, state park information and reservation, paying taxes and penalties etc. (Cook, 2000). Here, a critical success factor of such services is the trust of the users (OECD,

2009). An investigation by Venkatesch, Chan, and Thong (2012) showed that the most important attributes for transactional services are usability and security provision. In this pillar included are the financial and non-financial transactions available via the government website and it is divided into six parts: (1) online forms for filling in, (2) taxes, (3) penalties and (4) online payment for other fees, (5) library services (to extend the due date for borrowed books or to request loan-cards), (6) a personalized portal for the residents.

The fourth pillar, the interoperability (or integration), requires the ability of multiple governmental and non-governmental organizations to share and integrate information across their organizational boundaries (Pardo, Nam & Burke, 2011). The importance of interoperability was stressed above all by Gascó (2010), who pointed out many obstacles to respond to the emergence of new citizen demands, which cannot be covered by only one organization. In this pillar the horizontal and vertical integration are included. The vertical integration is achieved when it is possible to reach information and services from other agencies, departments or offices from just one website, i.e. the integration with libraries or museums. Horizontal integration is achieved when there is one database or intranet for all departments, and all departments use standardized software.

The fifth and last pillar is the participation. The so-called e-Participation enhances democracy and includes services like political surveys, political discussion forums or voting (Fietkiewicz et al., 2017). According to Saebo, Rose and Molka-Danielsen (2009), some of the challenges of e-Participation services are an identification of major user groups and addressing them during the development or the involvement of certain politicians and administrators in the participation process and a clear definition of their roles. Also, one critical aspect is the development of the user competence (technological literacy etc.). The last pillar includes services making it possible for citizens to leave feedback, make a complaint, or participate in an opinion survey. It is divided into four parts: (1) online surveys, (2) forums and electronic voting platforms, (3) participating online in local government meetings, and (4) online elections.

The investigated Qatar's e-Government website is called Hukoomi (Fig. 4.2). According to the website's statement, its role is "to make government information and services more efficient and effective in order to be accessible to all citizens, residents, visitors and businesses" as well as to raise "awareness on the public programs, events, news and government initiatives, in line with the objectives defined by Qatar e-Government 2020 Strategy" (Hukoomi, 2017). The first version of e-Government website was launched in 2003, followed by launching of new integrated e-Government in 2008 and constant improvements. In 2015 a Hukoomi mobile application was launched.

About ten years ago, Eggers (2005) introduced the term "Government 2.0" for the modified and increased utilization of new technologies by governments that is said to transform the government-citizen relationship. To make the interaction between these two parties more open and social, to improve their communication and to design user-centered governmental processes (Meijer, Koops, Pieterson, Overman, & ten Tije, 2012), governments join inter alia various social media

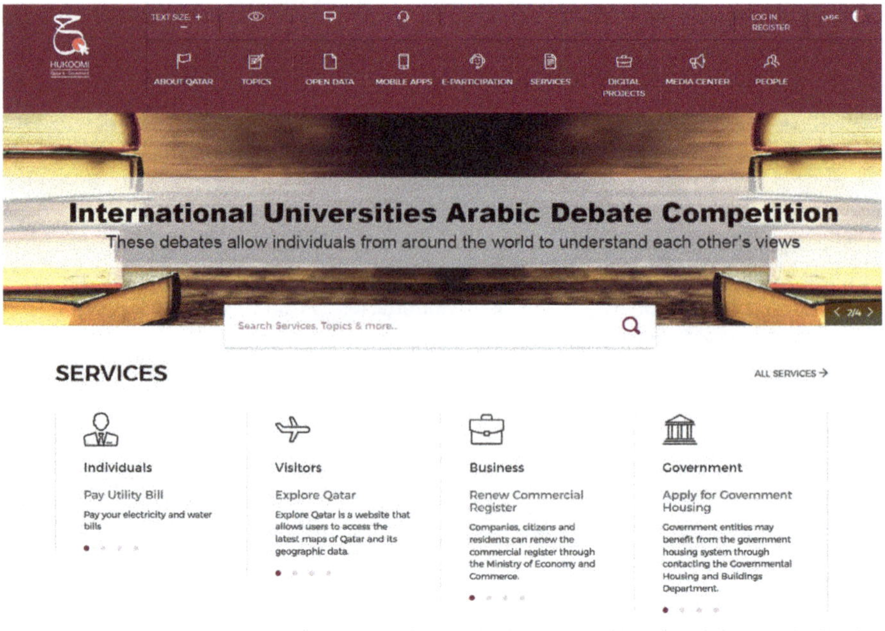

Fig. 4.2 The investigated governmental portal Hukoomi. *Source* http://portal.www.gov.qa

platforms. The use of these platforms leads to a cost reduction and to some possibly more important things: the trust and participation of the citizens achieved by transparency and openness of the government and, thus, a satisfied customer (Warkentin, Gefen, Pavlou, & Rose, 2002; Bertot, Jaeger, & Grimes, 2010). On the free market, the representation on social media platforms proved itself to be a common and important tool for marketing activities and many progressive governmental organizations started following this trend (Mangold & Faulds, 2009; Hartmann, Mainka, & Peters, 2013). In this way, the governments go online, communicate and cooperate with the citizens and invite them to participate in city-related events by trying "to reach their citizens where they are; namely on Facebook, Twitter, YouTube, etc." (Hartmann, Mainka, & Peters, 2013, p. 161).

Since we already know that many Qatari authorities are more or less active on different social media channels, we mainly focus our Government 2.0 investigation on their qualitative and quantitative output. Basis for this analysis are two lists of overall 35 authorities (Table 4.2) retrieved in December 2015 from the official Qatar e-Government website. Only the Qatar News Agency was omitted, because its output has no relation to the other institutions. Instead, we added the accounts of the current Qatari Emir as the most important political authority.

Table 4.2 List of the 35 analyzed authorities

Civil aviation authority	Communications regulatory authority
General authority of customs	Hamad Medical Corporation
Hukoomi—Qatar eGovernment	Ministry of Administrative Development
Ministry of Culture, Arts and Heritage	Ministry of Development Planning and Statistics
Ministry of Economy and Commerce	Ministry of Endowments and Islamic Affairs
Ministry of Energy and Industry	Ministry of Environment
Ministry of Finance	Ministry of Foreign Affairs
Ministry of ICT	Ministry of Interior
Ministry of Justice	Ministry of Labor and Social Affairs
Ministry of Municipality and Urban Planning	Ministry of Transport
Ministry of Youth and Sports	Childhood Culture Center
Mowasalat	Public Works Authority
Qatar Assistive Technology Center	Qatar Chamber
Qatar Financial Markets Authority	Qatar General Electricity and Water Corporation
Qatar Museums Authority	Qatar Postal Services Company
Qatar Tourism Authority	Regulatory Authority for Charitable Activities
Supreme Council of Education	Supreme Council of Health
The Qatari Emir Sheikh Tamim bin Hamad Al Thani	

Sources Hukoomi (2015a, b)

First, it had to be confirmed in which of the online services, Facebook[4] and Instagram,[5] the authorities have active accounts. We chose Facebook as one of the most frequently used services in informational world cities concerning governmental matters (Mainka, Hartmann, Stock, & Peters, 2014) and Instagram as a photo-sharing application that Qataris use more frequently (Dennis, Wood, & Martin, 2014). Since the information content of rather short Twitter posts seemed to be insufficient, we only focused on these two platforms. We retrieved the authorities' posts during a period of two weeks (January 2, 2016 to January 15, 2016). For every post, different indicators were recorded:

- number of followers (on Facebook and Instagram),
- date,
- number of likes (on Facebook and Instagram),
- number of comments (on Facebook and Instagram),
- number of shares (on Facebook) and
- language (Arabic and/or English).

[4]Facebook: https://www.facebook.com/.

[5]Instagram: https://www.instagram.com/.

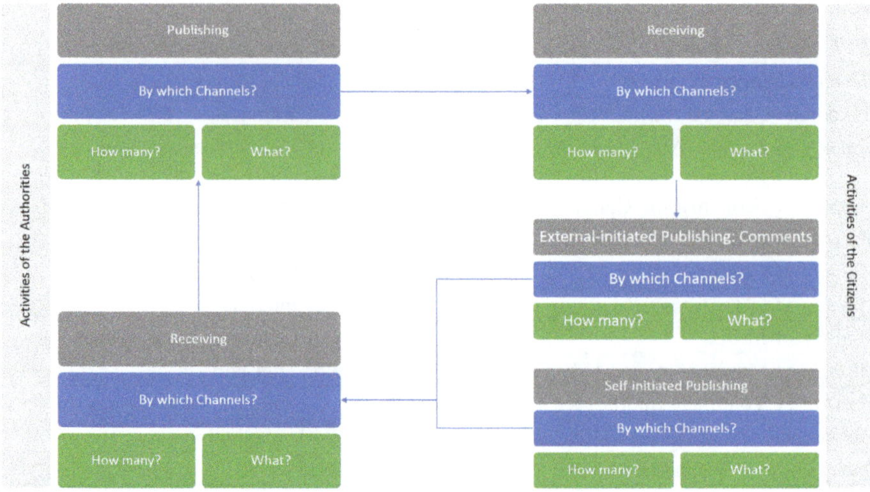

Fig. 4.3 Government 2.0 research framework

We looked up the number of followers on July 8th, 2016. The likes, comments and shares are evaluated for every single post fourteen days after its publishing date. For determining the language, we chose Arabic as the native language and English as the most spoken language in the region. These two languages are sufficient for classifying all posts. All data was collected manually.

The approach of analyzing posts is outlined in detail in Fig. 4.3. We analyzed both, the activities of the community (the authorities), and the activities of the citizens, each in terms of publishing and receiving. Concerning the citizens, we differentiated two kinds of publishing: (1) in the form of a comment on an institution's post and (2) in the form of a self-initiated post, which means that a citizen posts something on a government's social media site without a previous post by the institution. Overall, we evaluated the same data for the four types of publishing or receiving: the type of channel, the number and the content.

4.6 SWOT Analysis

The SWOT (acronym for strengths, weaknesses, opportunities and threats) analysis is very well suited method to arrange, group and organize already collected information. This common tool has a focus on both internal (strengths and weaknesses) and external (opportunities and threats) factors. Strengths and weaknesses include things that can be done by a firm or institution itself while the opportunities and threats refer to the supportive or unsupportive conditions of this firm's or institution's environment. By assessing the fit between these two parts, SWOT fulfills its role as a direction sign and "catalyst for the development of viable marketing plans" (Ferrell & Hartline, 2014, p. 85). Besides the analysis' simplicity, there are some other major benefits: (1) strategic planning costs can be reduced, (2) the use of a comprehensive information system can be used to fill structured information into the SWOT system directly and thus makes it very flexible, (3) it is possible to include and synthesize quantitative and qualitative information, and (4) the "collaboration and open information exchange between different functional areas" (Ferrell & Hartline, 2014, p. 87) goes on without any difficulties. If the analysis is conducted correctly and the strengths and opportunities are combined, SWOT is no longer an organizing tool, but rather provides the right strategy for the way into a prosperous future (Ferrell & Hartline, 2014).

In our study, the SWOT analysis will be used to clarify the strengths, weaknesses, opportunities and threats of Qatar's way into a knowledge society. We highlight what is already done right, what can be improved, how it could be implemented and what dangers Qatar should be aware of.

References

Abduljawad, H. (2015). Challenges in cultivating knowledge in university-industry-government partnerships: Qatar as a case study. *The Muslim World, 105,* 58–77.

Al-Kitbi, E. (2010). Where do GCC women stand in the development process? In C. Koch & L. Stenberg (Eds.), *The EU and the GCC: Challenges and prospects under the Swedish EU presidency* (pp. 95–103). Dubai, AE: Gulf Research Center.

Altvater-Mackensen, N., Balicki, G., Bestakowa, L., Bocatius, B., Braun, J., Brehmer, L., et al. (2005). Science and technology in the region: The output of regional science and technology, its strengths and its leading institutions. *Scientometrics, 63*(3), 463–529.

Becker, R. (2010). International branch campuses: New trends and directions. *International Higher Education, 58,* 3–5.

Bertot, J. C., Jaeger, P. T., & Grimes, J. M. (2010). Using ICTs to create a culture of transparency: E-government and social media as openness and anti-corruption tools for societies. *Government Information Quarterly, 27*(3), 264–271.

Böhme, G. (1997). The structure and prospects of knowledge society. *Social, 36*(3), 447–468.

Bothwell, E. (2016, January 14). *The world's most international universities 2016. Times higher education.* Retrieved from www.timeshighereducation.com.

Brewer, J. D. (2000). *Ethnography.* Buckingham, UK: Open University Press.

Cook, M. E. (2000). *What citizens want from E-Government.* Albany, NY: Center for Technology in Government.

Corbin, J., & Strauss, A. L. (1990). Grounded Theory research: Procedures, canons, and evaluative criteria. *Qualitative Sociology, 13*(1), 3–21.

Corbin, J., & Strauss, A. L. (2015). *Basics of qualitative research: Techniques and procedures for developing grounded theory* (Vol. 4). Thousand Oaks, CA: Sage.

Cronin, B. (2001). Bibliometrics and beyond: Some thoughts on web-based citation analysis. *Journal of Information Science, 27*(1), 1–7.

Dennis, E. E., Wood, R., & Martin, J. D. (2014). *Entertainment media use in the Middle East.* Doha, QA: Northwestern University.

Eggers, W. D. (2005). *Government 2.0: Using technology to improve education, cut red tape, reduce gridlock, and enhance democracy.* Plymouth, UK: Rowman & Littlefield.

Egghe, L. (2005). Expansion of the field of informetrics: Origins and consequences. *Information Processing and Management, 41*(6), 1311–1316.

Fenner, M. (2013). What can article-level metrics do for you? *PLoS Biology, 11*(10), e1001687.

Ferrell, O. C., & Hartline, M. D. (2014). *Marketing strategy: Text and cases* (6th ed.). Mason, OH: South Western Cengage Learning.

Fietkiewicz, K. J., Mainka, A., & Stock, W. G. (2017). eGovernment in cities of the knowledge society. An empirical investigation of smart cities' governmental websites. *Government Information Quarterly, 34*(1), 75–83.

Fietkiewicz, K. J., & Stock, W. G. (2014). Cityness and informativeness of the emerhing information cities in Japan. *Creative and Knowledge Society, 4*(1), 43–56.

Florida, R. (2002). The rise of the creative class: Why cities without gays and rock bands are losing the economic development race. *The Washington Monthly, 34*(5), 15–25.

Florida, R. (2003). Cities and the creative class. *City & Community, 2*(1), 3–19.

Florida, R. (2005). *Cities and the creative class*. New York, NY: Routledge.

Forster, N., Al Ali Ebrahim, A., & Alma Ibrahim, N. (2013). An exploratory study of work-life balance and work-family conflicts in the United Arab Emirates. *Skyline Business Journal, 9,* 34–42.

Gascó, M. (2010). Approaching e-government interoperability. *Social Science Computer Review, 30*(1), 3–6.

Gill, J. (2008, August 21). *Oiling the learning machine. Times higher education.* Retrieved from www.timeshighereducation.com.

Gisler, M. (2001). Electronic government – mehr als eine Website [Electronic government—more than just a website]. *DISP, 144,* 32–38.

Glaser, B. G. (1992). *Basics of grounded theory analysis: Emergence vs forcing*. Mill Valley, CA: Sociology Press.

Glaser, B. G., & Strauss, A. L. (2009 [1967]). *The discovery of grounded theory: Strategies for qualitative research.* Piscataway, NJ: Aldine Transaction.

Godin, B. (2007). Science, accounting and statistics: The input-output framework. *Research Policy, 36,* 1388–1403.

Godwin, S. M. (2006). Globalization, education and emiratisation: A study of the United Arab Emirates. *Electronic Journal of Information Systems in Developing Countries, 27*(1), 1–14.

Gremm, J., Barth, J., & Stock, W. G. (2015). Kuwait is the past, Dubai is the present, Doha is the future: Informational cities on the Arabian Gulf. *International Journal of Knowledge Society Research, 6*(2), 51–64.

Hall, P. (2005). The world's urban systems: A European perspective. *Global Urban Development, 1*(1), 1–12.

Hansson, F., Husted, K., & Vestergaard, J. (2005). Second generation science parks: From structural holes jockeys to social capital catalysts of the knowledge society. *Technovation, 25* (9), 1039–1049.

Hartmann, S., Mainka, A., & Peters, I. (2013). Government activities in social media. In P. Parycek & N. Edelmann (Eds.), *International conference for E-Democracy and Open Government 2013—CeDEM13* (pp. 173–186). Krems, AT: Edition Donau Universität Krems.

Haustein, S., Larivière, V., Thelwall, M., Amyot, D., & Peters, I. (2014). Tweets vs. mendeley readers: How do these two social media metrics differ? *IT—Information Technology, 56*(5), 207–2015.

Haustein, S., Sugimoto, C., & Larivière, V. (2015). Guest editorial: Social media metrics in scholarly communication. *Aslib Journal of Information Management, 67*(3).

Hay, I. (2010). *Qualitative research methods in human geography*. South Melbourne, Victoria: Oxford University Press.

Henkel, M. (2015). Educators of the information society: Information literacy instruction in Canadian informational cities. *Journal of Systemics, Cybernetics and Informatics, 13*(3), 22–27.

Henkel, M., & Stock, W. G. (2016). "We have big plans." Information literacy instruction in academic and public libraries in the United States of America. In *Proceedings of the 2nd International Conference on Library and Information Science, July 12–14, 2016, Kyoto, Japan* (pp. 159–175). Taipeh, TW: International Business Academics Consortium.

Hiller, J. & Bélanger, F. (2001). *Privacy strategies for electronic government.* Retrieved from http://www.businessofgovernment.org/sites/default/files/PrivacyStrategies.pdf.

Hoteit, L., Tuerpitz, A., & Luelsdorf, L. (2015). *Motivation and perseverance: The missing link to GCC knowledge economies*. Dubai, AE: The Boston Consulting Group.

Hukoomi. (2015a). *Government agencies.* Retrieved from www.gov.qa.

Hukoomi. (2015b). *Ministries*. Retrieved from www.gov.qa.

Hukoomi. (2017). *About Hukoomi.* Retrieved from http://portal.www.gov.qa/wps/portal/about-hukoomi.

Hvidt, M. (2015). *Transformation of the Arab Gulf economies into knowledge economies: Motivational issues related to the tertiary education sector.* Doha, QA: Arab Center for Research & Policy Studies.

ictQatar. (2014). *Qatar E-Government 2020 strategy*. Doha, QA: Ministry of Information and Communication Technology.

Johnston, N., & Williams, R. (2015). Skills and knowledge needs assessment of current and future library professionals in the state of Qatar. *Library Management, 36*(1/2), 86–98.

Kavanaugh, A. L., Fox, E. A., Sheetz, S. D., Yang, S., Li, L. T., Shoemaker, D. J., et al. (2012). Social media use by government: From the routine to the critical. *Government Information Quarterly, 29*(4), 470–491.

Kettl, T. (2002). *The transformation of governance*. Baltimore, MD: Johns Hopkins Press.

Kosior, A., Barth, J., Gremm, J., Mainka, A., & Stock, W. G. (2015). Imported expertise in world-class knowledge infrastructures: The problematic development of knowledge cities in the Gulf region. *Journal of Information Science Theory and Practice, 3*(3), 17–44.

Lee, S. Y., Florida, R., & Acs, Z. (2004). Creativity and entrepreneurship: A regional analysis of new firm formation. *Regional Studies, 38*(8), 879–891.

Maasen, S., & Weingart, P. (2005). *Democratization of expertise? Exploring novel forms of scientific advice in political decision-making*. Dordrecht, NL: Springer.

Mainka, A., Hartmann, S., Stock, W. G., & Peters, I. (2014). Government and social media: A case study of 31 informational world cities. In *Proceedings of the 47th Hawaii International Conference on System Sciences* (pp. 1715–1724). Washington, DC: IEEE Computer Society.

Malinowski, B. (1992). *Argonauts of the Western Pacific*. New York, NY: E. P. Dutton & Co.

Mangold, W. G., & Faulds, D. J. (2009). Social media: The new hybrid element of the promotion mix. *Business Horizons, 52*(4), 357–365.

Mansourian, Y. (2006). Adoption of grounded theory in LIS research. *New Library World, 108* (1228/1229), 386–402.

Mason, J. (2002). *Qualitative researching* (Vol. 2). London, UK: Sage.

Meijer, A. J., Koops, B.-J., Pieterson, W., Overman, S., & ten Tije, S. (2012). Government 2.0: Key challenges to its realization. *Electronic Journal of e-Government, 10*(1), 59–69.

Millen, D. R. (2000). Rapid ethnography: Time deepening strategies for HCI field research. In *Proceedings of the 3rd Conference on Designing Interactive Systems: Processes, Practices, Methods, and Techniques* (pp. 280–286). New York, NY: ACM.

Monfaredzadeh, T., & Krueger, R. (2015). Investigating social factors of sustainability in a smart city. *Procedia Engineering, 118,* 1112–1118.

Moini, J. S., Bikson, T. K., Neu, C. R., & DeSisto, L. (2009). *The reform of Qatar University.* Santa Monica, CA: RAND Corporation.

Moon, M. (2002). The evolution of e-Government among municipalities: Rhetoric or reality? *Public Administration Review, 62*(4), 424–433.

OECD. (2009). *Rethinking e-Government services. User-centred approaches.* Retrieved from www.oecd.org.

Pardo, T. A., Nam, T., & Burke, G. B. (2011). E-Government interoperability: Interaction of policy, management, and technology dimensions. *Social Science Computer Review, 30*(1), 7–23.

Priem, J., Groth, P., & Taraborelli, D. (2012). *The altmetrics collection. PLOS ONE, 7*(11), e48753.

Priem, J., Taraborelli, D., Groth, P., & Neylon, C. (2011). *Altmetrics: A Manifesto*. Retrieved from www.altmetrics.org.

Puppim de Oliveira, J. A., Doll, N. H., Balaban, O., Jiang, P., Dreyfus, M., Suwa, A., et al. (2013). Green economy and governance in cities: Assessing good governance in key urban economic processes. *Journal of Cleaner Production, 58,* 138–152.

QF. (2013). *Qatar foundation strategic plan 2013–2023*. Doha, QA: Qatar Foundation.

Rao, K. (2000). Free zones in the Middle East: Development patterns and future potential. In B. Hoekman & H. Kheir-El-Din (Eds.), *Trade policy developments in the Middle East and North Africa* (pp. 245–264). Washington, DC: The World Bank.

Saebo, O., Rose, J., & Molka-Danielsen, J. (2009). eParticipation: Designing and managing political discussion forums. *Social Science Computer Review, 28*(4), 403–426.

Sharma, S., & Palvia, S. (2010). E-government and e-governance. Definitions/domain framework and status around the world. In *5th International Conference on EGovernance* (pp. 1–12). New York, NY: Foundations of E-Government.

Sriramesh, K., & Rivera-Sánchez, M. (2006). E-government in a corporatist, communitarian society. The case of Singapore. *New Media & Society, 8*(5), 707–730.

Stock, W. G. (2011). Informational cities: Analysis and construction of cities in the knowledge society. *Journal of the American Society for Information Science and Technology, 62*(5), 963–986.

Stock, W. G., & Weber, S. (2006). Facets of informetrics. *Information – Wissenschaft und Praxis, 57*(8), 385–389.

Tadros, M. E. (2015). *The Arab Gulf states and the knowledge economy: Challenges and opportunities*. Washington, DC: Arab Gulf States Institute.

Tague-Sutcliffe, J. (1992). An introduction to informetrics. *Information Processing and Management, 28*(1), 1–3.

Ulmer, J. (2009). The knowledge society: Text and orality in Qatar. *The International Journal of the Book, 6*(1), 109–113.

Venkatesh, V., Chan, F. K. Y., & Thong, J. Y. L. (2012). Designing e-government services: Key service attributes and citizens' preference structures. *Journal of Operations Management, 30* (1–2), 116–133.

Vora, N. (2014). Between global citizenship and qatarization: Negotiating Qatar's new knowledge economy within American branch campuses. *Ethnic and Racial Studies, 37*(12), 2243–2260.

Warkentin, M., Gefen, D., Pavlou, P. A., & Rose, G. M. (2002). Encouraging citizen adoption of e-government by building trust. *Electronic Markets, 12*(3), 157–162.

Wiseman, A. W., & Alromi, N. H. (2003). The intersection of traditional and modern institutions in Gulf states: A contextual analysis of educational opportunities and outcomes in Iran and Kuwait. *Compare, 33*(2), 207–234.

Witte, S. (2010). Gulf state branch campuses: Global student recruitment. *International Higher Education, 58,* 5–6.

Yigitcanlar, T. (2010). Informational city. In R. Hutchison (Ed.), *Encyclopedia of urban studies* (Vol. 1, pp. 392–395). New York, NY: Sage.

Part III
Connection

Chapter 5
Diversification of Knowledge-Based Industries and Foreign Investment

Abstract The state of Qatar is in the minds of the people a wealth country that is able to rely on its oil and gas reserves. Because the reserves shrink little by little, the country faces the task of disengaging from this dependence in the nearest future. First steps were made by developing the tourism and MICE sector in a quite successful way. However, it is still a long way and a huge challenge for the country. The time will tell if they meet this challenge or if they decide for another option: the investment in foreign projects or companies.

Keywords Qatar · Industries · Diversification · Knowledge-based industries
Oil and gas industry · Information industry · Renewable energy industry
MICE industry · Tourism industry · Foreign investment

At the present, Qatar's economy is heavily specialized in hydrocarbon industries. Over 50% of the country's GDP are generated by those sectors—with the gas sector as the crucial part. Being aware of resources that slow to a trickle, and the need to quit the financial reliance on "the 'rent' revenues from the oil industry" (Reiche, 2010, p. 2396), to diversify the economy and with it to redistribute the GDP amounts, Qatar makes after-oil plans for the implementation of sustainable knowledge-based and competitive industries (Höselbarth, 2010; QF, 2013). That all industries depend to some—but not to the same—extent on knowledge inputs (OECD, 1999), is already recognized by the government and Qatar Foundation, who is the main actor during the implementation (QF, 2016). In combination with the required human resources, Qatar is in a position to stimulate the economic growth, which in turn might create demand for high-skilled jobs (QF, 2013). Until now, the country is rather far from this goal: Knowledge-intensive sectors—that generally refer "to those industries which are relatively intensive in their inputs of technology and/or human capital" (OECD, 1999, p. 18) and include among others chemical and pharmaceutical manufacturing, scientific research and development (R&D) services, telecommunications, software publishers as well as data processing—contribute only 3–4% to Qatar's GDP. In comparison, other benchmarks like Dallas or Seattle that started the enhancement of the share of GDP of

J. Gremm et al., *Transitioning Towards a Knowledge Society*,
https://doi.org/10.1007/978-3-319-71195-9_5

knowledge-based industries 20–35 years before, achieve nowadays values of 10% or higher. The Qatari government is a firm believer in the implementation of such a progress in the country, however, they expected it to proceed significantly faster. The strategical management of the nation's R&D investments together with the interaction of the support of existing and new knowledge-intensive industries as well as the buildup of a culture of innovation and entrepreneurship shall make for an advanced economic diversification in a minimum of time (QF, 2013). In all GCC states it is clear, that "diversifying economies away from oil sector was one goal of economic policies" (Almutairi, 2016, p. 84). We are going to describe some examples of Qatar's way of economic diversification:

- Oil and gas industry,
- Information industry,
- Renewable energy industry,
- MICE industry (meetings, incentives, conventions, and events),
- Tourism industry,
- Other promising industries, and
- Foreign investment.

5.1 Oil and Gas Industry

Nowadays, the existing industries in Qatar can be divided generally into those that relate on gas and oil and those that exist alongside, whereby "[t]he oil & gas sector has been, and continues to be, the main driver of Qatar's economy" (QF, 2013, p. 39). The industries are on a top level, have highest standards, but are often dominated by foreign companies, like for example Shell (Interview partner 9). Some non-oil and gas industries emerged because they were needed in the oil and gas sector. Much of the expansion of Qatar's non-oil and gas industries "would not have occurred without hydrocarbons. Construction, utilities and downstream transport services (margins from the shipment of LNG) all grew in response to the needs of a booming hydrocarbon sector. Growth in the petrochemical sector and in some energy-intensive activities was made possible by the availability of cheap feedstock" (Ibrahim & Harrigan, 2012, p. 14).

Only with the high public income from oil and especially gas, huge investments in knowledge infrastructures (including K-12 education, Qatar University, Education City, Sidra, etc.) became realizable (Weber, 2014). But what will happen, when the international prizes for oil and gas decrease significantly? "Ironically, given that the long-term goal of research for the Gulf states is to free themselves from oil dependence, the drop of the oil prize has reduced the amount of money available to fund science" (Sarant, 2016, p. S7).

For Ibrahim and Harrigan (2012, p. 20) it is clear, "for looking at potential opportunities for diversification, it makes good sense to start with hydrocarbons". However, is that really an option for long term economic development? Instead of being satisfied and feeling safe, oil-producing nations are suffering, because the stocks of oil and gas decline and the countries' future in this particular field is uncertain. The "oil production has severely declined in recent years, falling from 845,000 to 709,000 barrels of oil per day (bopd) between 2007 and 2014" (TOGY, 2015, p. 7). Furthermore, in other countries just other techniques (such as electro technique) are used (Interview partner 1). However, QF sets itself to help maximizing and safeguarding the long-term value of Qatar's hydrocarbon reserves by developing "new technologies to enhance recovery and processing" (QF, 2013, p. 39) of these resources, expanding the global market and supporting the upgrade to higher-value products. Qatar Petroleum—and its liquefied natural gas sub companies including Qatargas, RasGas and different LNG terminals—undertake the supervisor task, drive the Qatari energy industry and play the most significant role in restructuring the sector. Until now, the company's activities were streamlined, the international presence was expanded and the industry was diversified "by becoming the largest producer of gas-to-liquids in the world" (TOGY, 2015, p. 10). Despite the decrease of the oil reserves, the LNG companies in Qatar are comfortably off. The fact that for years and even today the share of oil and gas in the nominal GDP is over 50% results in the take home message that not the end of oil is the main factor destroying the economy, but rather the mass of oil that destroyed the economy in previous times (Interview partner 1).

If the locals neither change their attitude, nor support and accept the country's intentions that bank on other industries, Qatar will be out of the global picture real soon from now. This has already happened to other big institutions that did not follow new trends and there is possibly "no better example than Kodak of the importance of coming up with new ways of doing business" (Hamm & Symonds, 2006, 5th paragraph). Once an icon of technology innovation, they are now "fighting to recover from a tech revolution that is strangling its core business" (Hamm & Symonds, 2006, 5th paragraph). The company fails in the approach of finding an adequate replacement of its previous major revenue source—the film business. Relying on the sale of digital cameras and not on another innovation was wrong, because the best cameras' profit margins would have never been able to replace the profits of the film business. It is inevitable to find an alternative that corresponds in its margin to the previous business or industry—if this is not the case, the company is not able to survive economically. Kodak recognized the problem late, reacted slowly and took the wrong innovation path. Qatar has to prevent that the same happens to the country by making plans, collecting supporters and changing the nations' mind.

5.2 Information Industry

To reconstruct the nation's economy, Qatar and the other GCC states try to diversify in a broader context and not only in the hydrocarbon industry itself. They know that other technologies will take over the market. Especially the information and communication technology industries were sponsored to a great extent (Interview partners 1 & 4).

About 550 local, regional and international ICT companies are currently operating in Qatar. They provide ICT products and services, primarily "system integration, software development and consulting, and hardware maintenance" (MOTC, 2016b, p. 12). Additionally, "retail and wholesale (42% combined) are the other significant business activities of the ICT enterprises in Qatar" (MOTC, 2016b, p. 13). Thereby, most are rather small companies with fewer than fifty employees (84%).

To keep pushing the development of this sector forward, the Supreme Council of Information and Communication Technology (ictQatar) was established in 2004 to improve and promote the ICT development and to make this sector more vibrant and advanced. They introduce ICT plans, policies and programs as well as projects and initiatives, develop the next generation of (ICT) infrastructure, secure and enhance their efficiency and for this reason change the human, social, economic and environmental progress of the county (Manochehri, Al-Esmail, & Ashrafi, 2012; MOTC, 2016a). In 2010, additionally the Qatar Computing Research Institute (QCRI) was established "to build Qatar's innovation and technology capacity" (QCRI, 2016, 2nd paragraph) by "doing […] world-class multidisciplinary computing research" (QCRI, 2016, 3rd paragraph). All that with focus on the needs of Qatar and its National Vision, but also the needs of the rest of the region as well as the world. They analyze big data, conduct studies and frequently publish results that are of the nation's interest. The institution, acting under the umbrella of QF, collaborates with important international partners, like Google, Microsoft or the World Bank, and with some major local companies or institutions, like Al Jazeera, Qatar Statistics Authority and Qatar University (QCRI, 2016; Interview partner 4). Together they try to pave the way into a diversified knowledge economy.

However, since the companies themselves do not really invest in R&D—where the growth of the ICT sector nevertheless depends on—this part of the economy has to rely on the government's support in terms of the development of new products and services (MOTC, 2016b). Then again, with the additional help of QF's direct ICT investments in ICT-based companies, like the IT services and solutions provider Meeza (2016), the communication company Vodafone (Vodafone Limited, 2016) and the service provider Gulf Bridge International (GBI, 2016), an essential infrastructure for knowledge-intensive industries is created so that the country's plan can probably be put into action (QF, 2013). That the government gets involved in the continuous development is shown by the fact that overall about one-fourth of all ICT enterprises in Qatar realized more than ten projects in collaboration with and for the government between the years 2012 and 2014. So, "[a]ll types and sizes of ICT enterprises in Qatar are benefitting from the large number of government projects in Qatar" (MOTC, 2016b, p. 14). Despite the large support, the government is its own worst enemy because its policies are too hard, the registration process is too complex and an approval takes too much time. In some of the other GCC countries, these things are not as big a problem.

We found some approaches to integrate citizens in development projects of companies. In the "Smart Qatar Zone" of QITCOM 2017 (an event in Qatar discussing smart technologies), a hackathon competition happened. Young computer programmers and software developers from high schools and universities competed to development software solutions to real problems in companies.

5.3 Renewable Energy Industry

Is there any option to construct a green sustainable Qatar or at least a sustainable Doha? In many countries of the world, the environment is naturally green; however, in Qatar, there is desert. If Qatar or Doha indeed wants to become green, sea water from the Gulf has to be desalinated. Huge energy resources are needed for this water supply.

QF supports the renewable energy industry. Qatar plans to generate about 2% of its electricity by 2020 from solar power and even 20% by 2030. This should yield natural gas savings of 0.24 Mtoe (millions of tons of oil equivalent) in 2020 and 2.13 Mtoes in 2030 (Martín-Pomares et al., 2017, p. 1244). This recent investment goes to Qatar Solar Technologies, a solar energy company that has set itself the goal to become a world-leading provider (QF, 2013; QSTec, 2016). They construct the first solar power plant in Qatar that shall open in the year 2016 and shall produce up to 15 MW of energy. This will be the first step of Qatar's goal "to introduce the renewable energy technologies for power generation & water production technologies" (Kahramaa, 2014, p. 46) in the country and to produce overall 200 MW of solar energy to power therewith 66,000 homes. The government appreciates especially the state-own Advanced Metering Infrastructure (AMI) pilot project of KAHRAMAA, the General Electricity and Water Corporation of Qatar (Kahramaa, 2014; Interview partner 1).

Of course, Qatar has "a tremendous potential for development and deployment of solar power generation, due to the high solar irradiation levels and the availability of land" (Javed, Wubulikasimu, Figgis, & Guo, 2017, p. 123). However, in Qatar the environmental conditions are not easy to manage. There are sandstorms and dust which hinder the usefulness of photovoltaic panels. The dust leads to cement-like layers which are firmly bound to the glass surface of the photovoltaic modules and are tough to remove (Javed et al., 2017, p. 132). Dust storms reduce the output of solar devices. So an optimal cleaning approach of solar panels is needed.

In the year 2014, about 10,000 water and 15,000 electricity customers living in the two districts West Bay and Al-Sadd are among others equipped with Smart Meters, Data concentrators, Head End Systems (HES) and Meter Data Management Systems (MDMS). With the technology of Smart Meters it becomes possible to record the consumption in detail—compared to outdated electro mechanical meters. They report to a centralized MDMS and are stored there "to analyze the detailed consumption data and power quality data for better forecasting and optimization of energy use which helps to achieve energy saving targets and reduce carbon footprint" (Kahramaa, 2014, p. 30). The from now on feasible monitoring of consumption details through web portals allows the customers to plan and better regulate their usage. Until now, these technologies should cover the whole state, but whether this plan was successful is not clearly determinable.

Apart from that, the overall impression is that Doha is not green yet. The country tries to improve this, but as long as gas is available, truly smart projects will not happen—simply because gas is cheaper, it is a risk that they do not want to run and the necessity for a change does not exist. The government does not want to ignore the topic, but it sidelines the implementation of the plans because it does not want to invest all of its money. It should be added that locals do not pay for water, gas and electricity. They can consume as much as they want, so that in their view there is no reason to become green (Interview partners 1, 4, & 8). In terms of sustainability, an enthusiastic education team within Education City fathers different initiatives to start changing the minds of their students: "recycling and related awareness events in the student dorms; on-campus student jobs promoting sustainability; and environmental education events and annual fairs on campus" (Luomi, Crist, Alam, & Shakir, 2013, p. 4).

There are few projects of urban sustainability. Two "eco-icons" in Qatar (Koch, 2014, p. 1127) are Msheireb and Lusail. Msheireb is an urban regeneration project in Doha which aims to contribute to the environment. Lusail City (few kilometers north of Doha) is a city built from the scratch, giving accommodation for more than 200,000 inhabitants. "The Qatari media and advocates of the project frequently frame Lusail as the country's first sustainable city" (Koch, 2014, p. 1131). However, until today, most of the nationwide projects are just in process of planning. "The growing amount of green rhetoric in Qatar is not always matched by green practices" (Koch, 2014, p. 1120).

5.4 MICE Industry

The opening of Qatar National Convention Center (QNCC) in 2011 and the fact that the country hosts the 2022 FIFA World Cup necessitate that "Qatar will now have to invest enormous resources in new infrastructure" including "hotels, restaurants, stadiums and other sports facilities, railways, shopping centres and offices" (Trofimovskaya & Kirby, 2012, p. 4). With these improvements, the country tries to become more popular and caters for a more successful and more profitable development of the MICE tourism (Trofimovskaya & Kirby, 2012; Interview partner 4).

In the beginning of its first year, the convention center brought in 19.7 million U.S. dollar to Qatar and made a great leap forward until the end of the year 2012 (77.65 million U.S. dollar). This year has been successful especially due to UN's Framework Convention on Climate Change—"one of the largest conferences in the world" that "brought 10,000 delegates to QNCC" (OBG, 2014, p. 238). Maybe, not

Fig. 5.1 Qatar National Convention Center. The design is inspired by the Sidra tree that "symbolizes Qatar Foundation's three key pillars of education, science and research and community development" (Trofimovskaya & Kirby, 2012, p. 6). *Photo* Tamara Heck (2016; personal communication)

all of the conferences are very producing, but for the country's tourism industry it is very helpful to organize them (Interview partner 7). The prestigious building of QNCC (Fig. 5.1) near Education City innovates the four aspects: sustainability, technology, education and luxury (Trofimovskaya & Kirby, 2012).

The smaller Doha Exhibition and Convention Center (DECC), located in Doha's West Bay area, is a second important venue of Qatar wherewith the government tries to bring business people to Qatar (OBG, 2014; Interview partner 1; Fig. 5.2).

Thus, the basis seems to be existent, but the organization is in many cases in short supply. Interview partner 1 reports that one day he was not able to go home after visiting a friend because all the streets were overcrowded and closed. Hence, the organizers have to work on this infrastructure problem. The solution could be the planned railway network—with "151 km (94 miles) of line and 48 stations" (Hutzell, el Samahy, & Himes, 2015, p. 89)—which will range from the city center of Doha to Education City on the outskirts (Shaaban & Radwan, 2014). Spurred by the imminent host of the World Cup, the effort on the railway project has been ramped up (Hutzell et al., 2015).

Fig. 5.2 Doha Exhibition and Convention Center. *Photo* Tamara Heck (2016; personal communication)

5.5 Tourism Industry

The Qatari tourism sector is mainly promoted by Qatar Tourism Authority that introduced a sector strategy, which will be realized—with the help of different ministries, public entities, policy advocacy bodies and private stakeholders—until 2030. The strategy is another attempt to reduce the country's "reliance on its hydrocarbon resources, while also placing it on the world tourism map and helping promote and perpetuate its people's culture, values, and traditions" (QTA, 2014, p. 5).

Today, the Qatari tourism sector mainly attracts very rich tourists, which primarily come from Saudi Arabia or other GCC states (Interview partners 9 & 13). With the construction of the artificial island that is called Banana Island, a kind of Maldives for the well-heeled Arabian population was created. Going there costs about 50 U.S. dollars, and if you even want to stay a night, you have to pay around 400 U.S. dollars per night. However, the rich tourists are not being put off because of the prices. About 70–80% of the rooms are booked out during the season. To stay on the beach is only allowed when you checked in into the hotel—the Banana Island Resort Doha by Anantara—because the beach is not public, which is actually welcomed by the conservative Arabian customers. The attractions for people who

Fig. 5.3 A look inside the cozy Katara Cultural Village

come from far away are in contrast very limited. They can only visit cultural attractions, like the Museum of Islamic Art, the souk or the cultural village Katara (Fig. 5.3), or enjoy the few scenic beauties, like the corniche, the desert (Fig. 5.4) or the mangrove forest, whereby the range of opportunities gets wider in contrast to the past (Interview partners 1, 4, & 7).

Apart from that, the nature is rather charmless and there is not much to see and to do so that a few days are enough to experience everything and most of the tourists leave rapidly or stay mostly inside their hotels during their whole stay. Furthermore, there is another reason why non-Arabian tourists, if they come to Qatar, stay in their hotel: Inside the hotel they can do whatever they want, outside they are rather limited. Beaches are rather used as picnic areas as opposed to Western common behavior pattern as bathing facilities, maybe because national women should not wear bikinis or swimsuits. As a result, people from non-Arabian countries often experience a restriction of their movement. Same holds true for consuming alcohol. This is only possible in particular high-cost hotels, which could lead to a problem during the FIFA World Cup in 2022. Other regional tourist destinations, like Dubai, are on the contrary more socially open and have actually more tourist attractions so that European, American or Australian tourists indeed prefer to go to that place. Qatar tries to keep pace with Dubai, but the development of new attractions makes just a slow progress, while the choice of luxury four or five star hotels is already satisfactory. Besides the Western vacationer who is used to open behavior, the

Fig. 5.4 A luxury five-star desert camp in the desert of Qatar

Arabian withdrawn mentality likes even now the Qatari privacy and feels more comfortable there than in Dubai or Bahrain (Interview partners 1, 7, & 8).

Beside business tourists, wealthy Arab tourists and families that already make vacation in Qatar—whose numbers should still be raised—there are three longer-staying groups that shall be attracted: wealthy world travelers that look for unique and authentic destinations, authentic discoverers that are interested in exploring new destinations and cultures as well as Arab budget-conscious tourists. To remain competitive to the other GCC destinations, Qatar banks primarily on providing its culture, developing the urban infrastructure, hosting sport events, making better use of sea, beach and nature blessings and providing health and wellness facilities. Nowadays, the host of the FIFA World Cup 2022 is obviously the mostly noticed part, because the best event is nothing without viewers (QTA, 2014). Besides the planned soccer world cup, Qatar is recognized as "a hub for international sporting events" (Musinguzi, 2016, p. 265).

A first step of bringing people to the country is done by implementing the state-owned airline Qatar Airways that supports the two-day stopover idea (Interview partners 10 & 11). Besides the important diversification of the economy, the goals of the strategy implementation are clearly defined: “create more jobs, improve the country’s infrastructure, expand the opportunities for family-friendly leisure and recreation, enhance the ability to safeguard the country’s natural resources and wildlife, and, most of all, uphold and promote Qatari values and traditions” (QTA, 2014, p. 15).

In an empirical study on Qatar as a travel destination, British respondents of a questionnaire see Qatar as a country of low risk; however, they suggest that the country lacks appeal. “The presence of safety does not result in Qatar winning the hearts and minds of the British public” (Morakabati, Beavis, & Fletcher, 2014, p. 430). The partly negative press on the football world championship, on former relations to the Muslim Brotherhood and on the working conditions of foreign workers in Qatar did not help to create a positive image of Qatar (Samuel-Azran, Yarchi, Galily, & Tamir, 2016). Morakabati et al. (2014, p. 431) suggest “to promote a positive destination image” and to create a destination “that is appealing to leisure visitors”, especially to visitors from Western countries.

5.6 Other Promising Industries

There are many other fields Qatar could build up to generate further incomes: an agricultural industry by building farms with tomatoes, other vegetables and fruits, a sports industry to support events like the FIFA World Cup 2022 as well as to enhance the inhabitants’ sportive activity and the further expansion of the media industry in cooperation with, for example, the news network Al Jazeera, to establish a vibrant film industry that operates globally, are just some examples (Interview partners 1, 4, 5, & 13; Fig. 5.5).

Furthermore, there are plans to establish industrial zones where the sponsoring process does not take effect (Interview partner 4). So there seem to be huge plans “but you do not see that something happens” (Interview partner 13). Instead, many foreign talents head for home. Unsurprisingly, when regarding the fact that there is almost no flexibility in changing jobs. If one wants to do this, he would have to leave the country for a while—about two years—and could then return to take up a new job (Interview partners 1 & 4). New ideas, expressed by expats, are in most cases dashed rapidly, because the provider of the idea needs a sponsor and if the sponsor says no, the idea provider is outvoted. This is one reason why people do not bring new ideas to or create them in Qatar. Indeed, Qataris want to be independent of oil and gas, but the incentive to establish a really productive industry is nominal. The country is exactly the opposite: Qatar is to one hundred percent a consuming society—not at least due to the implementation of Education City which “run[s] the risk of making the GCC countries consumers, rather than producers” (Mackenzie Smith, 2008, p. 19)—and the majority is not prepared to start working (Interview partners 1, 9, & 10).

Fig. 5.5 Qatar encourages the people to make new discoveries

5.7 Foreign Investments

Thus, it seems to be a chance to have a look abroad and to bank and distribute the nation's money in foreign investments. "Another means through which Qatar seeks to enhance its influence regionally and internationally is by international investments through its sovereign wealth fund" (Kamrava, 2015, p. 96). Qatar's wealth fund is the Qatar Investment Authority (QIA), which was established in 2006. QIA's portfolio "is intended to create long-term value for generations to come. It is therefore not subject to conventional short-term performance measures or tactical portfolio optimisation. Our approach enables us to execute investments in a skillful and timely manner, deploying our flexibility and dynamism to take advantage of long-term market developments and trends" (QIA, 2016). Main assets include "Volkswagen, Barclays, Canary Wharf, Harrods, Credit Suisse, Heathrow, Glencore, Tiffany & Co., Total" (QIA, 2016). These days, QIA holds 13% in the German automobile manufacturer Volkswagen AG's preferred shares and 17% in their ordinary stock as well as 8.2 percent of the commodity trade company Glencore (Stein, 2015). Furthermore, the country purchased the popular department store Harrods and several skyscrapers (a greater part of the skyline) in London. Qatar seems to be the largest property owner in London, UK. Additionally, QIA is involved in the French organization Institut Français (Alraouf, 2012; Interview partner 4).

QIA invests heavily in Europe; however, it is also active in other world regions. Kamrava (2015, p. 99f.) reports, "the 2011 Arab Spring provided another investment opportunity for Qatar, with the Qatar Investment Authority investing some $543.8 million in Egypt, for example, and creating 4000 jobs in the country's hard-hit economy." Via Hassad Food, QIA has diverse vertical agricultural investments "in Qatar, Australia, Pakistan and Oman with potential future investments in Asia, Africa, Europe and North and South America" (Hassad Foods, 2015). One of the major consequences of Hassad's investments is "to enhance Qatar's global financial status and, especially insofar other developing countries are concerned, their dependence on continued investments and revenue flows from the small Sheikdom" (Kamrava, 2015, p. 102). With successful foreign investments, parts of Qatar's wealth are and will be in future secured by companies and their employees working for Qatar outsides the borders of the country.

5.8 Summarizing Qatar's Diversification Strategies

Today Qatar has two strategies: (1) to develop an own diversified economy and (2) to invest in foreign businesses. Which way the country will follow in the next years and which way is the more promising one seems not to be fixed, because there are some industries that make progress, like the MICE and tourism industry, but in contrast, the overall impression of our interview partners is that diversification does not proceed very quickly. To support the process of developing country-owned resources, Carnegie Mellon University in Education City started the Executive and Professional Education program that offers different courses in which the faculty shares its world-class expertise with 755 professionals—415 from the government and 340 from the private sector. The yearlong program covers different disciplines like quality, innovation, service and leadership as well as business and e-business strategy and tactic (QF, 2015). However, always in the fore is the goal of "meeting the needs of, and securing a high standard of living for, all its people both for the present and for the future" (GSDP, 2008, p. 11). Whether it is possible to realize this purpose with own resources seems to depend on the way of implementation, the openness of the government for new ideas and companies as well as the courage of leaving the traditional, once secure oil and gas-centric economic path. Otherwise, they may indeed just have the opportunity to rely on foreign companies and their profits.

Having supporters—in this term the government and Qatar Foundation—of new sectors is nevertheless not the only thing that matters. What does the country need additionally? Can creative heads that develop new ideas, constructs and strategies help? For government workers that are in their routine jobs for years it might be difficult to put new strategies into practice. Is it possible to recruit such creative people to come to Qatar or to search for them amongst the inhabitants? Will the country accept their help? Without a concrete plan, it does not make sense to invest in one or two trends that maybe do not have any long-term effect. Which are the subject areas and sectors that are important to develop a successful and diversified economy?

References

Almutairi, H. (2016). Economic diversification in GCC economies: A heaven for investors. *International Journal of Economics and Finance, 8*(4), 84–92.

Alraouf, A. A. (2012). Dohaization: Constructing a new urbanity brand amid Gulf and Middle Eastern cities. Retrieved from www.analchemyofarchitecture.blogspot.com.

GBI. (2016). Gulf Bridge International. Retrieved from www.gbiinc.com.

GSDP. (2008). *Qatar National Vision 2030*. Doha, QA: General Secretariat for Development Planning.

Hamm, S., & Symonds, W. C. (2006, November 27). Mistakes made on the road to innovation. *Bloomberg*. Retrieved from www.bloomberg.com.

Hassad Foods. (2015). *Securing the Future*. Doha, QA: Hassad Foods. Online: http://www.hassad.com/English/AboutUs/Pages/default.aspx.

Höselbarth, F. (2010). *The education revolution in the Gulf*. Hildesheim, DE: Georg Olms.

Hutzell, K., el Samahy, R., & Himes, A. (2015). Inexhaustible ambition: Two eras of planning in Doha, Qatar. *Architectural Design, 85*(1), 80–91.

Ibrahim, I., & Harrigan, F. (2012). Qatar's economy: Past, present and future. *QScience Connect 2012, 9*, 24.

Javed, W., Wubulikasimu, Y., Figgis, B., & Guo, B. (2017). Characterization of dust accumulated on photovoltaic panels in Doha, Qatar. *Solar Energy, 142,* 123–135.

Kahramaa, (2014). *Connecting—Sustainability Report 2014*. Doha, QA: Qatar General Electricity & Water Corporation.

Kamrava, M. (2015). *Qatar: Small state, big politics*. Ithaca, NY: Cornell University Press.

Koch, N. (2014). "Building glass refrigerators in the desert": Discourses of urban sustainability and nation building in Qatar. *Urban Geography, 35*(8), 1118–1139.

Luomi, M., Crist, J. T., Alam, B., & Shakir, M. B. (2013). Environmental sustainability in Qatar's Education City: Strategies, initiatives and education. *QScience Connect*, 1–11.

Mackenzie Smith, P. (2008). Introduction. In C. Davidson, & P. Mackenzie Smith (Eds.), *Higher education in the Gulf states: Shaping economies, politics and culture* (pp. 9–22). London, UK: SAQI.

Manochehri, N., Al-Esmail, R., & Ashrafi, R. (2012). Examining the impact of information and communication technologies (ICT) on enterprise practices: A preliminary perspective from Qatar. *The Electronic Journal of Information Systems in Developing Countries, 51*(3), 1–16.

Martín-Pomares, L., Martínez, D., Polo, J., Perez-Astudillo, D., & Bachour, D. (2017). Analysis of long-term solar potential for electricity generation in Qatar. *Renewable and Sustainable Energy Reviews, 73,* 1231–1246.

Meeza. (2016). *Careers*. Retrieved from www.meeza.net.

Morakabati, Y., Beavis, J., & Fletcher, J. (2014). Planning for a Qatar without oil: Tourism and economic diversification, a battle of perceptions. *Tourism Planning and Development, 11*(4), 415–434.

MOTC. (2016a). *About us*. Retrieved from www.ictqatar.qa.

MOTC. (2016b). *Qatar's ICT Landscape 2016 Business*. Doha, QA: Ministry of Transport & Communications.

Musinguzi, D. (2016). Trends in tourism research on Qatar: A review of journal publications. *Tourism Management Perspectives, 20,* 265–268.

OBG. (2014). *The Report Qatar 2014.* Oxford, UK: Oxford Business Group.
OECD. (1999). *OECD Science, technology and industry scoreboard 1999 benchmarking knowledge-based economies.* Paris, FR: OECD Publishing.
QCRI. (2016). *About QCRI.* Retrieved from www.qcri.com.
QF, (2013). *Qatar Foundation Strategic Plan 2013–2023.* Doha, QA: Qatar Foundation.
QF, (2015). CMU-Q executive education program hones leadership skills. *The Foundation, 80,* 3.
QF. (2016). *About.* Retrieved from www.qf.org.qa.
QIA. (2016). *The investment policy and the list of main investments.* Doha, QA: Qatar Investment Authority. Online: http://www.qia.qa/Investments/InvestmentsApproach.aspx.
QSTec. (2016). *Who we are.* Retrieved from www.qstec.com.
QTA. (2014). *Qatar National Tourism Sector Strategy 2030/Q&A.* Doha, QA: Qatar Tourism Authority.
Reiche, D. (2010). Energy policies of Gulf Cooperation Council (GCC) countries: Possibilities and limitations of ecological modernization in rentier states. *Energy Policy, 38,* 2395–2403.
Samuel-Azran, T., Yarchi, M., Galily, Y., & Tamir, I. (2016). Promoting terror or sport? The case of Qatar's international image. *American Behavioral Scientist, 60*(9), 1101–1115.
Sarant, L. (2016). The Middle East. An end to oil dependency. *Nature, 537*(7618), S6–S7.
Shaaban, K., & Radwan, E. (2014). Rebuilding the transportation system in the city of Doha. *Journal of Traffic and Logistics Engineering, 2*(3), 241–247.
Stein, C. (2015, September 28). Qatar fund loses $5.9 billion on holdings in VW, Glencore. *Bloomberg.* Retrieved from www.bloomberg.com.
TOGY. (2015). *The Oil and Gas Year Qatar 2015.* Dubai, AE: The Oil and Gas Year.
Trofimovskaya, I., & Kirby, A. (2012). *MICE Destination Qatar Case Study: Qatar National Convention Centre.* London, UK: European Business School London.
Vodafone Limited. (2016). *Who we are.* Retrieved from www.vodafone.com.
Weber, A. S. (2014). Linking education to creating a knowledge society: Qatar's investment in the education sector. *Handbook of Research on Higher Education in the MENA Region: Policy and Practice* (pp. 52–73). IGI Global: Hershey, PA.

Chapter 6
Creativity and the Knowledge Society

Abstract Qatar is a country where creativity and innovation has until now not really been a prevalent topic. Only with the plan to become an urban region of the knowledge society did these concepts gain attention. These days, most of the creatives in Qatar are still from overseas. However, taking a look at Qatar's publications and patents, it is visible that the country is trying to encourage inventions and the exchange of knowledge.

Keywords Qatar · Creativity · Knowledge society · Innovation
Openness · Tolerance · Science—technology—medicine (STM)
Scientific output · Publications · Qatar University · Texas A&M University Qatar
Weill Cornell Medical College Qatar · Patents · Everyday knowledge
Knowledge workers

Nowadays creativity is a fundamental resource for the growth and development of cities, whole regions and nations. The creativity of individuals is reflected in innovations, in new developments in commerce and industry and in changes in community and civilization. Economic growth related to creativity is, according to Florida (2005), based on three T's: The first T stands for technology, which is a product of human creativity and crucial for economic success, the second T means talent measured by creative occupations and the third T is tolerance which is a must have for every place that wants to attract technology and talent. Tolerant places are characterized by "diversity—an openness to all kinds of people, no matter their gender, race, nationality, sexual orientation, or just plain geekiness" (Florida, 2014, p. 9). Those who act in creative occupations belong to the "creative class". Its members are "engaged in science and engineering, research and development, and the technology-based industries, in arts, music, culture, and aesthetic and design work, or in the knowledge-based professions of health care, finance, and law" (Florida, 2005, p. 3). All of these occupations have at least one thing in common: They focus on employees that work mainly with their minds (Florida, 2014). To exploit the creativity that slumbers in the citizens, it is necessary to establish research infrastructures to provide the best possibilities for doing research

J. Gremm et al., *Transitioning Towards a Knowledge Society*,
https://doi.org/10.1007/978-3-319-71195-9_6

(Alraouf, 2011). To reach economic growth through creativity, it is also of great importance to give people the opportunity to initially create and exchange knowledge by establishing public spaces for communication as well as by providing information and communication technologies (David & Foray, 2002).

Besides job opportunities, salaries and safety, there are only few motivational factors for international creatives (including scientists) to come and to stay in Qatar. Attractive areas are scarce; national Qatari visit Aspire Zone, Katara, Souk Waqif, The Pearl, the Corniche and the West Bay area—that is all (Salama, Wiedmann, Thierstein, & Al Ghatam, 2016, p. 297; Salama & Wiedmann, 2016). Trips to the desert are very interesting, but only one or two times. While there are indeed nice beaches in Qatar, nationals, especially Qatari women, prefer not to go swimming in the sea. To wear a bikini is taboo. For expats being non-Moslem permanent residents, there is some limited alcohol available legally (via the Qatar Distribution Company or in international hotels); however, drinking is not allowed at public places. Pork products are available for non-Moslems in only one shop in the entire country. "The missing development of integrated urban centres has thus led to a high dependency on shopping malls and thus the experience of Doha as a highly replaceable city by a majority of migrants. Place making and the creation of functioning neighbourhoods are thus further key challenges to attract the creative class to settle in Doha long term and to establish a diversified economy" (Salama et al., 2016, p. 322).

The National Vision of Qatar proves that the state is well aware of the fact that nowadays the transformation of societies into knowledge-based ones is necessary to maintain a running and successful economy. With its plan for the future, the country has set the goal to develop this kind of society by establishing advanced educational and health systems and the corresponding infrastructures, and by engaging in scientific research and intellectual activity. Additionally, the Gulf state aims at promoting openness and tolerance towards other cultures and at encouraging and supporting innovation and creativity (GSDP, 2008). First steps to realize these goals were already taken, initiated by the founding of Qatar Foundation, which is an important driving force behind Qatar's 2030 Vision (QF, 2016).

6.1 Creativity and Innovation

According to our interviewees, the creativity in Qatar is rather limited. When talking about creatives in Qatar, one has to distinguish between local and expat creatives. Since local creatives are rare, the country tries to allure and employ highly educated and creative expats (Interview partners 1 & 5).

> New city urban development and its spatial qualities contribute to the global attractiveness of Doha for knowledge economy investments, firms and people. These urban developments fulfill the requirements of Knowledge workers coming to the city from literally every spot of the globe anticipating an attractive smooth quality of life which would foster their creativity and innovation (Alraouf, 2011, p. 56).

In the last few years, especially scientists and engineers were given the opportunities to come to the country. Adequate infrastructures were established that encouraged them to create innovations and to share their knowledge with Qatar. While the royal family tries to attract creatives engaged in the fields of art, music and design to establish an art and cultural scene, this has been of little success until now (Interview partners 2, 6, 7, 9, & 10). Interviewee 3 is wary about Qatar's creative class. He is of the opinion that due to the process of Qatarization, the amount of creatives in Qatar is shrinking. Moreover, creative people face some problems to use their full potential of creativity because they need total freedom including freedom of expression to exploit their talents (Interview partner 1).

Probably due to missing creatives in Qatar, an innovative atmosphere is not really present in the country, yet. However, the room to become an innovative country is there, given the fact that Qatar funds a lot. Furthermore, the country is willing to discard and change things quickly and there are lots of opportunities in Qatar (Interview partners 1 & 12). The country's attitude is positive in this regard. It tries to create an innovative atmosphere as it supports and encourages innovations (Fig. 6.1). Interviewee 4 even says it tries to force them. The Qatar National Research Fund tries to encourage innovation and hopes that while expanding and diversifying Qatar's economy, the nation embraces innovation (QNRF, 2016). However, this holds only true for commercial ideas. When it comes to innovations

Fig. 6.1 With the help of Education City the country tries to promote creativity

in other fields, there is little support in Qatar. For non-Qataris, it is even more difficult to use their inventive talent, so that a lot of innovativeness stays hidden (Interview partners 1 & 6).

6.2 Openness and Tolerance

Regarding openness and tolerance in Qatar, the opinions of the inhabitants are divided. While some say there is no openness or tolerance at all (Interview partner 8), others say that Qatar is far too open and too tolerant (Interview partner 3) and still others say that while Qatar is a tolerant country, the adjective "open" cannot be used to describe it (Interview partners 1, 7, & 9). The aspect of a segregated society is observed by many of our interviewees. Locals and expats are virtually forced to coexist in Qatar, but they live apart from each other in different areas of the city and their lives barely intermingle (Interview partners 2 & 9). Interviewee 10 mentions that Qataris are scared of foreign infiltration. This is why they keep to themselves and spend their weekends in desert tents where they seek distance from expats while bonding with the family and other natives (Fig. 6.2).

Despite these habits, interview partner 10 is of the opinion that Qataris are very hospitable. Interview partner 12 thinks that in general Qataris are willing to be open

Fig. 6.2 Qataris often spend their time in the desert and drive around the desert

because “otherwise they would not try for knowledge society” and they would never bring as many expats into the country. However, he mentions that this does not hold for the whole country, but that there is certain resistance in parts of the society. It is those Qataris who say the country has become far too open and too tolerant. Especially the universities and other higher education institutions in Qatar are places for openness and tolerance. The whole complex of Education City is set up for tolerance and freedom (Interview partners 3, 12, & 13). It attracts students with different nationalities and cultures arriving from countries spread all over the world. More than 60 nationalities come together in Education City and everyone is accepted there. “It is a tolerant and respectful atmosphere” (Küster, 2010, p. 17). Qatar University also tries to familiarize students “with foreign cultures, practices, and views and to encourage students to approach foreign cultures with open minds, respect, and tolerance” (Moini, Bikson, Neu, & DeSisto, 2009, p. 25).

6.3 Scientists, Engineers and Physicians and Their Scientific Output

Where there is creativity and an innovative atmosphere, there are likely to be highly qualified workers, like scientists, engineers and physicians. If those places are open and tolerant towards different kinds of people, those experts arrive from countries all over the world to carry on their professions there.

As recently as a few years ago, most of the universities and other higher education institutions in Qatar were only teaching institutions and the scientists working there did not perform any research activities. By now, this has changed and most of the academics teach and perform research at the same time. Not only faculty has adopted research as a new task, students are also encouraged to carry out research and get included in the universities’ research projects (Interview partners 4, 6, & 9). At Qatar University, undergraduate students are expected to engage in research to gain experience just as faculty members (Moini et al., 2009). However, although the country and the universities are aware of the important benefits regarding research, teaching is still the dominant part in the academics’ weekly schedule. Our interviewees who have made those experiences themselves confirmed this proposition. Some of them mentioned that in Qatar it is difficult—sometimes even problematic—to perform research and teaching simultaneously. Students in Qatar need a lot of personal support. It is not uncommon that the academic staff supervises them outside of class hours, by repeating already discussed topics or helping them with assignments. Beside these additional favors, the time to perform research runs low (Interview partners 1, 7, & 13). While scientists working at universities are therefore constraint in performing research activities, there are institutions in Qatar that were built for scientists to do just research. By bringing lots of great researchers into the country, Qatar has proceeded to stuff those research institutions (Interview partner 12), like Qatar Computing Research

Institute for applied computing research (QCRI, 2016), Qatar Biomedical Research Institute (QBRI) for basic and applied medical research (QBRI, 2016) and Qatar Environment and Energy Research Institute (QEERI) for research in energy and water security (QEERI, 2016). "To attract researchers, the Qatar Foundation now offers state-of-the-art research facilities which are helping to keep world-class researchers and professors in Education City and develop knowledge sharing and development" (Küster, 2010, p. 17). So even though the research culture of the country still has to develop, the quality of the research that actually is conducted is already high (Interview partners 9 & 12).

> In the last 10 years Qatar has emerged as the country with the best citation footprint among GCC nations, although the nation has the lowest percentage of documents cited. In other words, when articles from Qatar-based institutions are cited, their citation dynamics surpass the GCC and world averages (Herciu, 2016, p. 3).

Whether scientists give political advice based on their knowledge acquired through research or not, strongly depends on connections. It is not a common practice of Qatar's scientists to give advice to politics and administrations. Some scientists do, but others do not. The ministries reach out to scientists sometimes, but not systematically. Very few scientists are working directly with the government. Normally, the political leaders have their own limited group of consultants working for them. Just in rare occasions, scientists working at universities are invited by the politicians to attend and contribute to their seminars (Interview partners 1, 4, & 7).

Since research activities increased in Qatar, the scientific output in form of publications has slowly started to become visible (Interview partner 7). Publication counts retrieved from Web of Science and Scopus also confirm this claim. Table 6.1 shows the number of Qatar's publications in Web of Science and Scopus in the last ten years from 2006 to 2015. Altogether, Scopus includes more of Qatar's publications than Web of Science. In every single year from 2006 to 2015, the amount of publications from the preceding year was surpassed. This holds for both scientific databases. In total, there are 10,741 publications of Qatar recorded in Web of Science, of which 8040 emerged in the last ten years. In the scientific

Table 6.1 Number of Qatar's publications (2006–2015). *Sources* Web of Science, July 4, 2016; Scopus, July 4, 2016

Year	Web of science	Scopus
2015	2308	3196
2014	1707	2798
2013	1178	1946
2012	800	1374
2011	564	1010
2010	446	835
2009	327	666
2008	300	569
2007	247	460
2006	163	312
TOTAL	8040	13,166

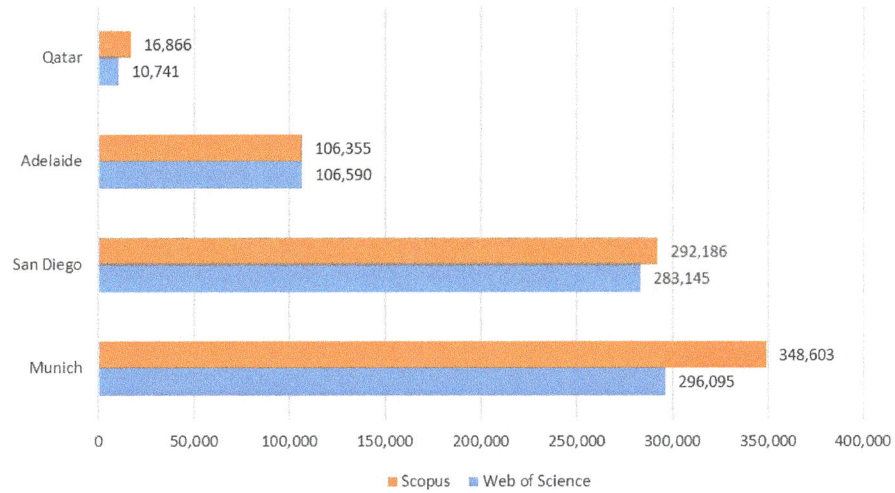

Fig. 6.3 Number of Qatar's publications in comparison to those of Adelaide, San Diego and Munich. Sources: Web of Science, July & August, 2016; Scopus, July & August, 2016

database Scopus, 16,866 publications from Qatar are listed. 13,166 of those have been published in the time from 2006 to 2015.

This at least small success with regard to the country's scientific output has been encouraged and strongly supported by the Qatar National Research Fund. However, when comparing these numbers to those of Adelaide, San Diego and Munich, cities that are comparable to Doha in population numbers, the publication numbers of Qatar are comparably small (Fig. 6.3). In Web of Science and in Scopus, the address where the fewest of publications come from is Qatar. The number of texts from Munich in Scopus is the largest in comparison to Qatar and the other two cities as well as in comparison to the publications of Munich registered in Web of Science. With a total of 348,603 the city has exactly 331,737 publications more registered in Scopus than Qatar. Of the three cities that are adduced for comparison, Adelaide has the fewest registered publications, in Scopus even less than in Web of Science. However, with a total of 106,355 publications registered in Scopus it still exceeds Qatar by 89,489 papers.

Qatar's research is specialized on the country's transfer into the knowledge society (Interview partner 7). According to our interview partners 1 and 5, the core themes of the publications are defined by the QNRF. It has prioritized four main topics on which research should focus, namely "energy and environment", "computer science and ICT", "health and life sciences" and "social sciences, arts and humanities" (QNRF, 2016). Web of Science and Scopus reveal the most discussed topics of Qatar's researchers. All of the frequently discussed topics are in line with the four pillars of the Qatar National Research Fund. The most discussed topic of Qatar's publications in Web of Science is "electrical and electronic engineering". This is the topic of exactly 720 publications. "Sport sciences" is the topic of 588 publications and 458 texts revolve around "chemical engineering". The next seven frequently used categories are titled "multidisciplinary materials science", "physical chemistry", "telecommunications", "general and internal medicine", "multidisciplinary chemistry, "energy and fuels" and "cardiac and cardiovascular systems". The ten most relevant categories in Scopus correspond topically to those in Web of Science. In Scopus the ten main topics are "medicine", "engineering", "computer science," "physics and astronomy", "biochemistry, genetics and molecular biology", "chemistry", "mathematics", "materials science", "social sciences" and "energy". Thereof, medicine is with a total of 5076 publications the most discussed topic by researchers in Qatar, followed by engineering (3828 publications) and computer science with 2920 publications.

When it comes to universities, some are more engaged in publishing than others, depending on their specialization and size. Interviewee 4 believes that Texas A&M University and Weill Cornell Medical College are much more active in publishing than, for example, Northwestern University due to their specializations in engineering and medicine. Interviewee 10 mentions that University College London has already produced good results in terms of publications in the short time since they started to operate in Education City. Data from Web of Science regarding the affiliation of the publishers confirm the assumptions of interviewee 4 that of all higher education institutions in Education City Texas A&M University and Weill Cornell Medical College are the universities that publish most of the articles. While Texas A&M University Qatar has published 1278 texts in the years from 2011 to 2015, Weill Cornell Medical College Qatar has published 825 articles in that time interval. These two universities have produced a large amount of publications in Qatar, but the number of publications in Web of Science where members of Qatar University are involved, is even higher, which might be due to the larger amount of faculty members in comparison to the branch universities. Qatar University published 1831 articles in the time from 2011 to 2015. When regarding the publications of the three institutions for each year individually, the annual increase becomes visible. Qatar University and Weill Cornell Medical College have succeeded in surpassing their research output in form of publications in every single year during the last five years. Texas A&M University has also increased its output in the time from 2011 to 2014 consistently, but in 2015, three publications less are recorded. In all examined years, Qatar University has published the most and Weill Cornell Medical College the fewest publications in comparison to the other institutions (Fig. 6.4).

Companies in Qatar that were attracted by the sudden wealth of the country, are constantly in need of qualified engineers and technical staff (Parsaei & Elbashir, 2014). The existing engineers in Qatar are mostly highly qualified expats that are bought from other parts of the world. Most of the engineers working in the country are architects and construction engineers, having nothing to do with teaching. Since most of the buildings in Qatar are established by expats, those foreign engineers play a decisive role for Qatar's economy (Interview partners 1, 9, & 10). The Texas A&M University in Qatar has "recognize[d] the profound need for highly skilled engineers to lead the next generation of discovery" (Parsaei & Elbashir, 2014,

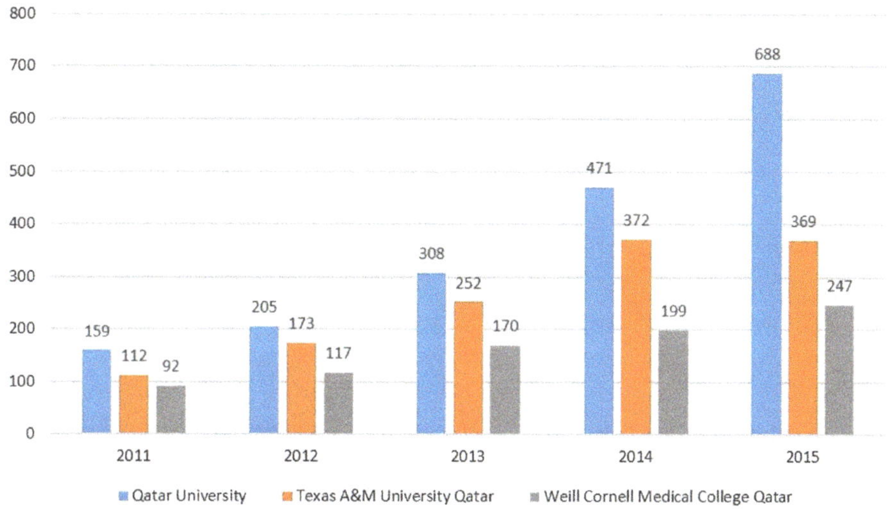

Fig. 6.4 Number of publications of Qatar University, Texas A&M University Qatar and Weill Cornell Medical College Qatar (2011–2015). *Source* Web of Science, July 26, 2016

p. 24.80.2). It has developed a model to provide students with the best engineering education possible. To develop their skills and guarantee their success in future workplaces, students are allowed to participate in research projects in fields relevant to the Qatar National Research Strategy and get encouraged to undertake internships and cooperative learning experiences.

"The virtual absence of a culture conducive to innovation and supportive of the spirit of research and invention in the Arab World could not be more evident than through a rundown of the number of patents issued to the region" (Madar Research & Development, 2014, p. 20). Due to the fact that still most of the engineers are foreigners staying in the country for only five or maximal ten years, they do not strive for applying patents in Qatar. They are mainly involved into practical work with the goal to construct high-class infrastructure (Interview partners 2 & 10). Interviewee 9 mentions that Qatar Foundation has applied a few patents, but all in all our interview partners are not informed about the existence or the origin of patents from Qatari companies or institutions. Interview partner 4 ventures the guess that some patents might come from Texas A&M University or Weill Cornell Medical College due to their specializations in engineering and medicine and interviewee 2 remarks that Qatar University might have applied some patents because they also have an engineering program, but both of them cannot say this for sure.

With the help of the scientific database Scopus we receive the number of patents that were applied by innovators of Qatar at patent offices all over the world. In July 2016, exactly 339 patents from Qatar are recorded in Scopus. Figure 6.5 shows the number of recorded patents in the last ten years. Most of them were published in the

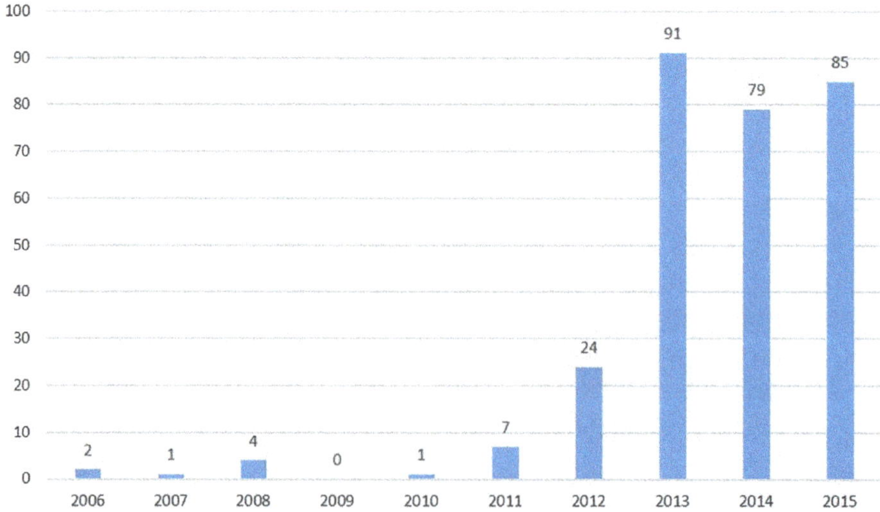

Fig. 6.5 Number of Qatar's patents (2006–2015). *Source* Scopus, July 04, 2016

year 2013, followed by 2015 and then 2014. In the years before 2012, Qatar did not apply or was not granted many patents.

Again, we adduce the cities Adelaide, San Diego and Munich to compare their patent numbers to those of Qatar (Fig. 6.6). The same as for the publications holds: Qatar has less patents than all three of the compared cities. When taking a look at

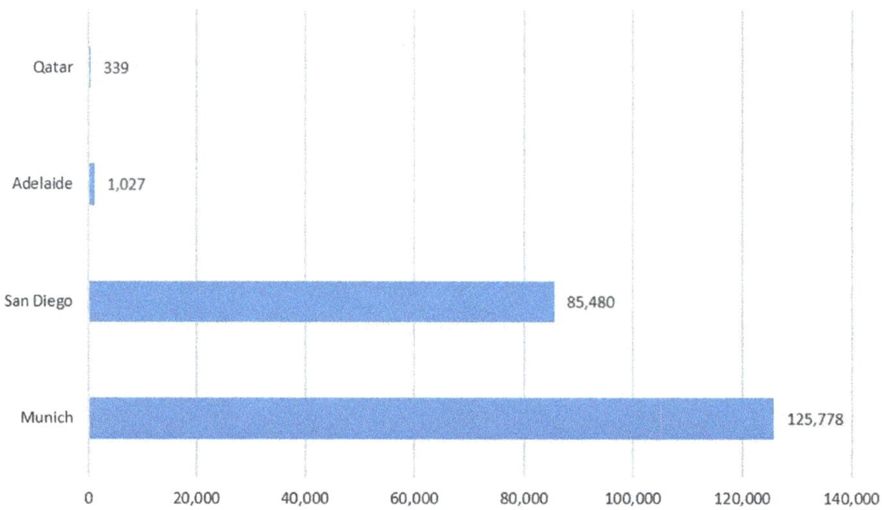

Fig. 6.6 Number of Qatar's patents compared to those of Adelaide, San Diego and Munich. Source: Scopus, July & August, 2016

Munich which has with a total of 125,778 the most publications in Scopus, altogether 125,439 more than Qatar, it becomes obvious that Qatar has still a long way to go.

Qatar is trying to build first class medical facilities with only the best staff from Western countries. This process has just started and the infrastructure is still under construction. Sidra, a hospital for women and children, is the current big medical project of the country. However, it is not completely put into service, yet. So, although some of the world's best physicians have already been brought into the country, they were not able to start practicing in the hospital. There is already good primary health care in the country, but when it comes to more complicated procedures, people are brought overseas for medical treatment (Interview partners 1, 8, 10, & 11). While talking about the quality of physicians in Qatar we asked our interviewees if they go to the local hospitals. Surprisingly most of them replied that they would fly home for medical treatment or go to Dubai instead. Only few of them said that they were perfectly fine with doing it in Qatar because the standard was alright (Interview partners 4 & 13). According to interviewees 4 and 12, medical education does exist. Qatar University has a medical degree program just like Hamad Bin Khalifa University and Weill Cornell Medical College. None of the interviewees comments on the quality of education of these universities. The hospitals in the region are, however, not only expected to give medical treatment, but also to conduct research. The Hamad Medical Corporation is already very good in doing research. They work on research projects with about 50 external experts. Sidra is planned to open as a hospital, but it is already designated that research will be added later on (Interview partners 7 & 13).

6.4 Exchange of Everyday Knowledge

The most important substance of a knowledge-based economy is the knowledge itself. Qualified knowledge workers and students possess a lot of relevant knowledge. One of the goals is to share and exchange the available knowledge, to extend the knowledge base and come up with new ideas. Knowledge can be exchanged face to face or with the help of information and communication technologies.

When talking about the exchange of everyday knowledge between the people, one has to differentiate between the habits of the younger versus that of the older generation (Fietkiewicz, Baran, Lins, & Stock, 2016) and between business affairs versus private matters. The way to communicate in Qatar has shifted. Things are no longer discussed only face to face, but through different means (Interview partners 1 & 6). Nearly everyone who uses the internet is engaged in social media or direct messaging (Dennis, Martin, & Wood, 2015). The majority of conversations are conducted via WhatsApp (Interview partner 12), which is used by all age groups (Dennis et al., 2015). The youth in Qatar is a big fan of using other social media platforms. They communicate via all new channels, preferably Instagram and Snapchat, but also Twitter, Facebook and YouNow (Interview partners 1 & 12).

Contrary to this, the older generation in the Gulf region is not as eager to use these platforms (Dennis et al., 2015). Whenever something new comes out, the younger generation is eager to use it, replacing the older channels by the new ones (Interview partner 7). Especially females in the Gulf region use the internet to socialize with their friends. Males, on the other hand, rather like to meet their friends in person to communicate with them face to face (Dennis et al., 2015). Nowadays, students run around "with one hand always on their mobile" (Interview partner 8). In Qatar, the coverage of cell phones is with 168 percent high—some people even own three smartphones. Some of our interviewees told us that the academic staff was encouraged to use new media for teaching. Especially some members of the older generation that strictly stick to traditions exchange knowledge personally. Once a week, they meet with family and friends outside of their main house in a tent called *majlis* to socialize. The head of the family is thereby always the host of the meeting. Mostly male visitors are received in these tents because it is uncommon for males and females to socialize together (Interview partners 4, 7, 8, & 12). However, if Qatar wants to become truly knowledge-based, it has to overcome some obstacles given by its history. The distribution of information in modern societies mainly happens via texts. Consequently, Qatar, where the primary means of communication is orality, has to shift to a text-oriented society (Ulmer, 2009). When it comes to business affairs, however, people in Qatar still prefer to communicate face to face. In companies and in universities, the important things are discussed in personal meetings. Even emails, representing the more classical form of digital media, are rarely responded (Interview partner 1).

6.5 Summarizing Qatar's Creativity and Knowledge Base

Qatar is well aware that its development towards a knowledge society in order to compete in the global economy is inevitable. In its National Vision 2030, the country recognizes that important issues need to be addressed. However, so far, the desirable atmosphere of openness and tolerance, which is a key to deal with further requirements of a knowledge society, is not fully present, yet. Qatar is multiply divided: On the one hand, there are Qataris as a minority in their own country and thus having a strong feeling of being overwhelmed by immigrants. On the other hand, there are expats, many of them well educated, by whose work the country benefits, but who live in constant fear of losing their job and being sent out of the country. The expats are yet again divided by their professions that range "from services and construction to management and science" (Powell, 2014, p. 258). However, even Qataris—some more concerned about sticking to culture and religion than others—are not all on the same page. Just recently, 14 sculptures, positioned in front of Sidra Medical and Research Center by order of Sheikha Al Mayassa, displaying different stadiums from conception to birth, were covered up because some of the Qatari nationals felt offended by the artwork (Interview partners 10 & 11). This state is problematic in a country where creativity and innovation only recently started receiving the appropriate attention. Highly qualified creative workers, like scientists, engineers and physicians, who are the primary driving force of innovation, are mostly brought in from other countries to help Qatar succeed. However, the country has recognized its need of well-educated workers and tries to educate experts in the required fields with the help of Western branch campuses. To exploit the talent of scientists, engineers and physicians, Qatar encourages and funds research in the areas in which they are engaged. For a long time, the creation of publications and patents was not a priority in Qatar. This has changed within the last few years. Even though the number of publications and patents is in general quite low, the recent increase in scientific output is visible. The rise of publication and patent numbers shows that Qatar is trying to become more innovative and that it has already made some considerable progress in the last few years. But it still has to face some obstacles: Traditionally, Qataris communicate orally rather than in written form. Even though students nowadays prefer to exchange knowledge via messaging services and social media, business affairs are principally discussed personally. Since in knowledge societies information and knowledge is mainly exchanged via texts, Qatar has to recognize and address this problem in order to use the existing knowledge exhaustingly.

Just as knowledge production by knowledge workers and its exchange as well as the general exchange of knowledge between citizens is an important aspect of a knowledge-based society, the communication of important knowledge and information from the government to the people, but also from the people to the government, is crucial for a running knowledge society. The examination of communication between the government and its citizens is analyzed in the following under the term e-Government. Has the concept of e-Government arrived in Qatar? How and by which channels do the political elites connect with the citizens? What about the citizens? Do they try to get in contact with the governmental institutions? How mature is the e-Government portal Hukoomi?

References

Alraouf, A. A. (2011). Assembling identity for an emerging urbanity: Doha, Qatar as a creative city. *Proceedings of The City: 2nd International Conference—Culture, Society, and Technology. Vancouver, British Columbia, Canada, 4–6 May 2011* (pp. 48–63).

David, P. A., & Foray, D. (2002). An introduction to the economy of the knowledge society. *International Social Science Journal, 171,* 1–22.

Dennis, E. E., Martin, J. D., & Wood, R. (2015). *Media use in the Middle East. A six-nation survey.* Doha, QA: Northwestern University.

Fietkiewicz, K. J., Baran, K. S., Lins, E., & Stock, W. G. (2016). Other times, other manners: How do different generations use social media? In 2016 Hawaii University International Conferences. Arts, Humanities, Social Sciences & Education, January 8–11. (2016). *Honolulu, Hawaii, Proceedings* (pp. 1–17). Honululu, HI: Hawaii University.

Florida, R. (2005). *Cities and the creative class.* New York, NY: Routledge.

Florida, R. (2014). *The rise of the creative class: Revisited.* New York, NY: Basic Books.

GSDP. (2008). *Qatar National Vision 2030*. Doha, QA: General Secretariat for Development Planning.

Herciu, I. (2016). *Qatar's top 500 scientific articles: Most cited versus most used research*. New York, NY: Thomson Reuters.

Küster, A. (2010). Qatar—a new beginning? The development of a collaborative research and education environment in Education City, Doha. Serials, *23*(1), 16–19.

Madar Research & Development. (2014). *Arab Knowledge Economy Report 2014*. Dubai, AE: Orient Planet.

Moini, J. S., Bikson, T. K., Neu, C. R., & DeSisto, L. (2009). *The reform of Qatar University*. Santa Monica, CA: RAND Corporation.

Parsaei, H. R., & Elbashir, N. (2014). A new vision for engineering education. *Proceedings of the 121st ASEE Annual Conference & Exposition, Indianapolis, IN, June 15–18, 2014* (p. 24.80 (8 Pages)). Washington, DC: American Society for Engineering Education.

Powell, J. J. (2014). University roots and branches between "Glocalisation" and "Mondialisation": Qatar's (inter)national universities. In A. W. Wiseman, N. H. Alromi, & S. Alshumrani (Eds.), *Education for a knowledge society in Arabian Gulf countries* (pp. 253–276). Bingley, UK: Emerald.

QBRI. (2016). *Vision*. Retrieved from www.qbri.org.qa.

QCRI. (2016). *Vision, mission and values*. Retrieved from www.qcri.com.

QEERI. (2016). *Overview*. Retrieved from www.qeeri.org.qa.

QF. (2016). *About*. Retrieved from www.qf.org.qa.

QNRF. (2016). *Vision & mission*. Retrieved from www.qnrf.org.

Salama, A. M., & Wiedmann, F. (2016). Perceiving urban liveability in an emerging migrant city. *Proceedings of the Institution of Civil Engineers—Urban Design and Planning, 169*(6), 268–287.

Salama, A. M., Wiedmann, F., Thierstein, A., & Al Ghatam, W. (2016). Knowledge economy as an initiator of sustainable urbanism in emerging metropolises: The case of Doha, Qatar. *ArchNet—International Journal of Architectural Research, 10*(1), 274–324.

Ulmer, J. (2009). The knowledge society: Text and orality in Qatar. *The International Journal of the Book, 6*(1), 109–113.

Chapter 7
E-Government

Abstract The launch of Hukoomi as Qatar's main e-Government portal has been one of the first essential steps to successfully implement the country's e-Government strategy. Some aspects like the service of paying fees online or voting on a non-political level are already available and used, but the expansion of further services, possibilities of user participation as well as the optimization of the website's structure in general are unavoidable if Qatar wants to keep up with the world's most mature regions regarding this topic. Also, the social media activity of Qatar's ministries and authorities is in just few cases fully accepted and used for a government-citizen communication. Still, most authorities need to catch up regarding this matter, too.

Keywords Qatar · E-government · Hukoomi · Governmental websites Maturity · Social media · Facebook · Instagram · Followers · Posts Likes · Shares · Comments · Qatar tourism authority · Qatar ministry of interior E-services

The Qatari e-Government was launched in July 2000 (EL-Haddadeh, Weerakkody, AL-Shafi, & Ali, 2010) "to e-enable all government" (Kostopoulos, 2004, p. 296). With this new opportunity to participate in governmental occurrences, Arab governments have to face the challenge of accepting changes—"the change in mindset that citizens must be in the heart of any development plan" (Al-Khouri, 2013, p. 13). Citizens have to become customers, to be a part in designing and developing the country and, hence, need channels to get in contact with the government. Only this way the government-citizen relationship can be established and the citizens can be satisfied (Al-Khouri, 2013). It is already recognized that e-Government has to obtain priority now and in the future, but also that there is much more work to achieve what has been remarked in the Qatar e-Government Strategy 2020 (ictQatar, 2014), which consists of three main purposes:

J. Gremm et al., *Transitioning Towards a Knowledge Society*,
https://doi.org/10.1007/978-3-319-71195-9_7

- to serve individuals and business by offering 100% of the governmental services online,
- to create efficiency by automating functions, modern applications and a well-developed ICT infrastructure,
- to develop a more open government with the ability to participate and offer a free access to open data.

The Qatari government promises that "[a]ll individuals and businesses will benefit from connecting online with Qatar's more open and efficient government" (ictQatar, 2014, p. 3). Customers will save time, have a better user experience and enhanced service quality. The government will be more productive, have freed-up resources and will share platforms, data and resources. The whole nation will benefit from more transparency finding expression in a better life for the citizens, an enhanced economic development, international recognition and transparent data access as well as participation.

To get an in-depth look into the actual advances and the country's e-Government status, we conducted two studies: (1) the measurement of the maturity of Qatar's official e-Government portal Hukoomi and (2) an investigation of Qatar's main governmental Facebook accounts—as one of the most governmentally used social media channel—and also of Qatar's main governmental Instagram presences—as "those that their citizens also use" (Mainka, Hartmann, Stock, & Peters, 2015, p. 251). We are going to analyze the development of the government-citizen relationship.

7.1 Maturity of Governmental Websites

Taken the five pillars catalogue, communication, transaction, integration and participation that are shown in Fig. 7.1 as a whole, Qatar reaches a little less than the half of the overall score.

That places the city/country in a global context in the last quarter of a ranking of the maturity of informational world cities that has been developed in the study by Fietkiewicz, Mainka and Stock (2017). Overall 31 informational cities all over the world were analyzed in their study, however, Doha, and respectively Qatar, were not included. But, with the results of our investigation it is possible to pin Qatar down somewhere between the 24th and 25th rank of the international study. The only city of the Gulf Region that participated in this worldwide comparison was Dubai, scoring 3.3 points less than Qatar, which is a rather comparable result. Regarding the maturity of e-Government, cities like Beijing, San Francisco and Paris are also comparable to Qatar. The same study indicates Barcelona as the top performer that reaches overall about 140 points more than Qatar.

When regarding the five pillars separately, we can probably reveal the weak spots of Qatar's e-Government presence Hukoomi. In the first pillar catalogue, also

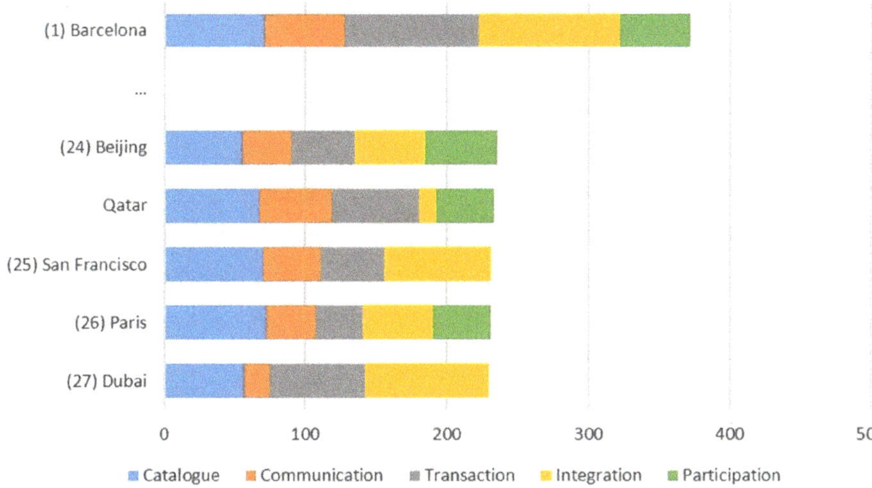

Fig. 7.1 Extract of the ranking of maturity of 31 e-Governments including the score of Qatar. *Source of foreign data* Fietkiewicz et al., (2017)

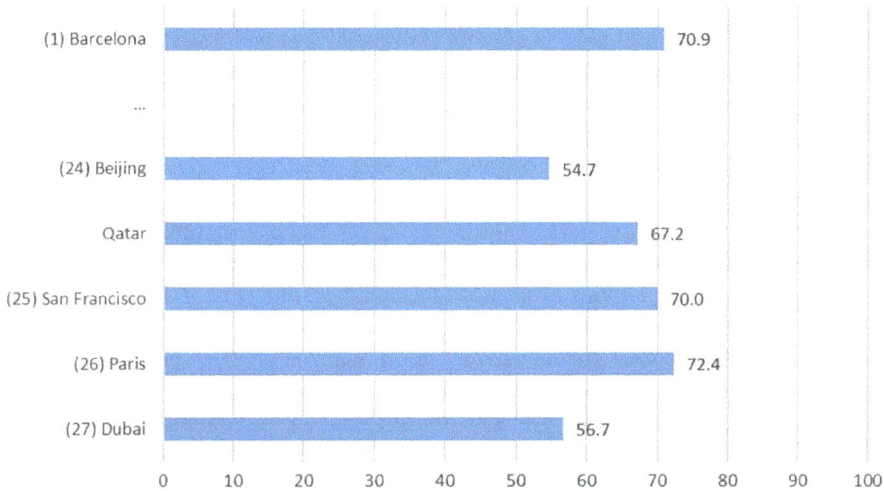

Fig. 7.2 Pillar one: Catalogue. Qatar reaches a good result. *Source of foreign data* Fietkiewicz et al., (2017)

called information dissemination, Qatar obtains about 67 points of overall 100 points and gets close to Barcelona's result (Fig. 7.2).

The results indicated that the users have sufficient access to basic information like, for example, about the healthcare system and medical care, about politics and services as well as customized information for residents, tourists or students. Indeed a lot of information can be obtained by mobile applications for Android and iOS,

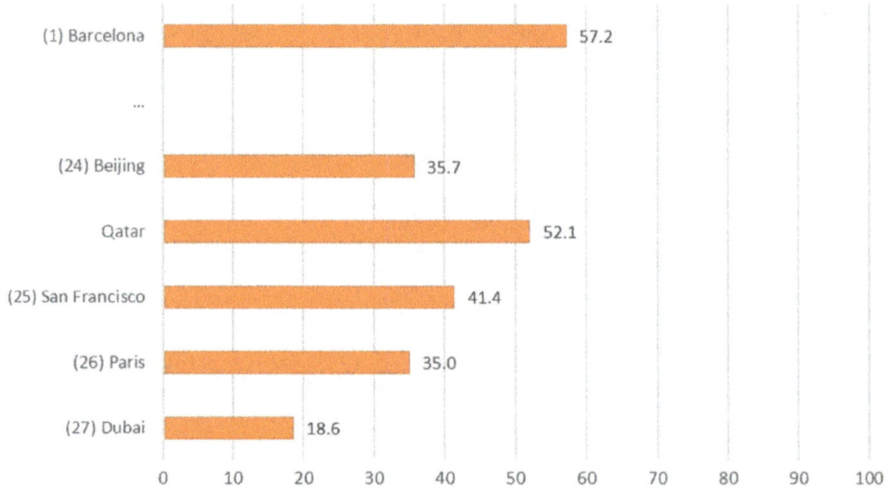

Fig. 7.3 Pillar two: Communication. Qatar reaches again a good value, this time with a greater distance to the comparative cities. *Source of foreign data* Fietkiewicz et al., (2017)

but the usage of the mobile version of the webpage is rather not satisfactory, since the content of the original page is notably shortened. The web texts are available in Arabic and English, what probably is sufficient for this region due to the fact that everyday conversation is conducted in English. For the immigrants speaking only broken English, the translation into their native language would be helpful for better orientation.

The second pillar includes the two-way communication between government and residents. Also in this part of the study Qatar reaches a comparatively good result. As a comparison, the regional rival Dubai achieves less than 20 points, what is approximately a little more of a third of Qatar's results (Fig. 7.3).

A wide spectrum of social media channels are used—analyzed more detailed in our e-Government social media investigation part—and it is possible to give feedback or to leave complaints. However, the reply to email requests leaves much to be desired. Even though we got a response to our message that we sent via the contact form, it occurred first after ten days. Therefore, for urgent questions the answering service could be too slow. In addition to it, there is no possibility to get emails instead of written mails via the website.

In the transaction pillar that includes financial and non-financial transactions, Qatar gets comparatively quite many points. Here, the wheat separates from the chaff, as the top performers get fairly 70 or even more points, and there are only few cities that reach very bad results in this category (Fig. 7.4).

Qatar is hold back by the library services. It is not possible to extend book loans or to apply for a card to borrow books from the library. Probably they outsource these processes to the library webpage, but in that case a hint as well as an external link would be helpful. Indeed, it is possible to pay fees online, but considering that

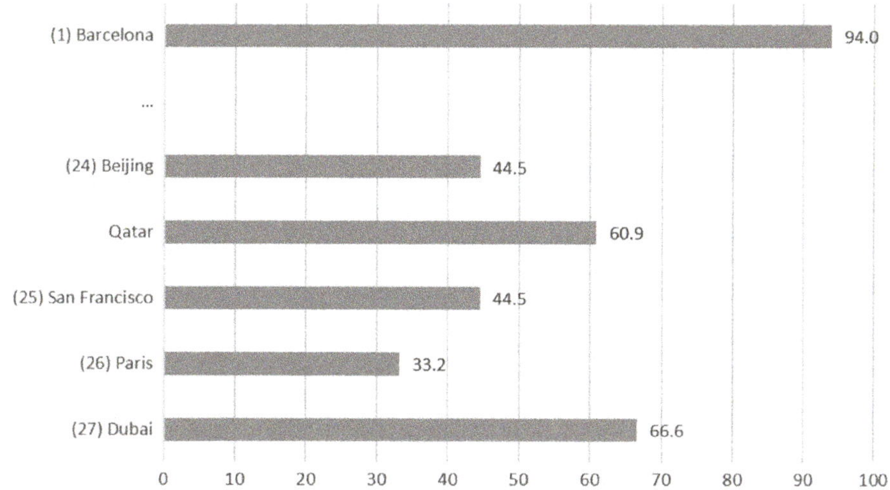

Fig. 7.4 Pillar three: Transaction. In the direct comparison Qatar makes a good score, but the result is little inferior to the one of the regional rival Dubai. *Source of foreign data* Fietkiewicz et al., (2017)

some services are offline or simply not accessible, Qatar could not get a full score. A personalizable profile can ease the handling of eServices by bookmarking the mostly used ones.

The fourth pillar—integration—is the one at which Qatar performs worst. No more than 12.5 points of overall 100 points are achieved (Fig. 7.5).

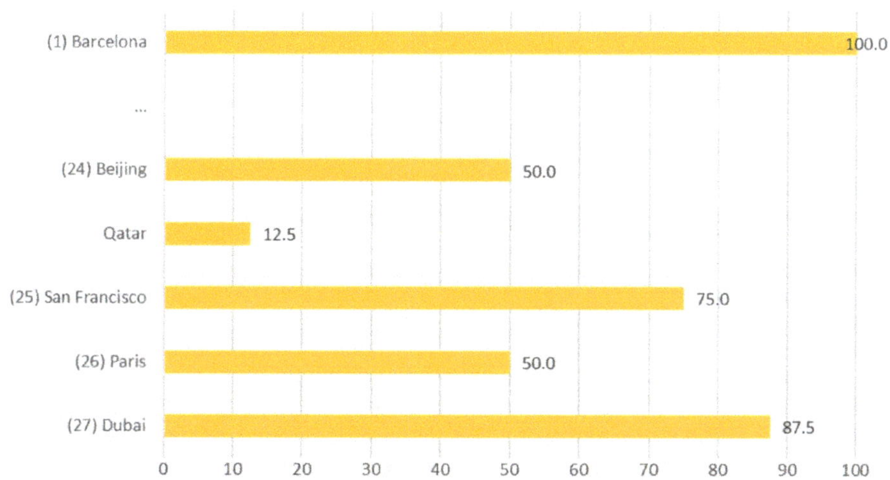

Fig. 7.5 Pillar four: Integration. Qatar reaches only few points. *Source of foreign data* Fietkiewicz et al., (2017)

Indeed, Hukoomi has an entry homepage, however, important links to webpages of, for example, museums and libraries or to information about how to register a car in Qatar are not available. To get this information, users have to take an in-depth look. The entry page is divided in information blocks for different user-groups including students, citizens and residents or visitors, what might be helpful in some cases. But the study shows that in most cases users have to click-through for quite some time to find the page of interest.

Lastly, we analyzed the participation possibilities offered on Qatar's e-Government presence Hukoomi. Compared to other cities, Qatar reaches a quite good result. No other city scored more than 50 points, whereas Qatar scores 40 points and, therefore, belongs to the high-scored regions. But, in comparison to the best possible result (here again 100 points), we can say that the citizens' participation is nowhere near adequate (Fig. 7.6).

Qatar seems to acknowledge the importance of e-Participation, especially in comparison to other e-Governments, but topics like the implementation of forums or platforms for asking questions, as well as the possibilities to participate in community meetings via the WWW or to vote online still have to be developed.

Qatar's e-Government, as we have seen, exhibits heterogeneous maturity levels for different areas. Regarding the pillars catalogue and communication, Hukoomi can match up with the very mature e-Government of Barcelona. Actual press releases, information on healthcare, politics and services are available and in many cases divided between individual boxes for different user-groups. The presence on many social media platforms also stands out. How successful they are in this field will be discussed later. What should Qatar reevaluate? Especially the method of integrating services into the portal should be considered. Finding services

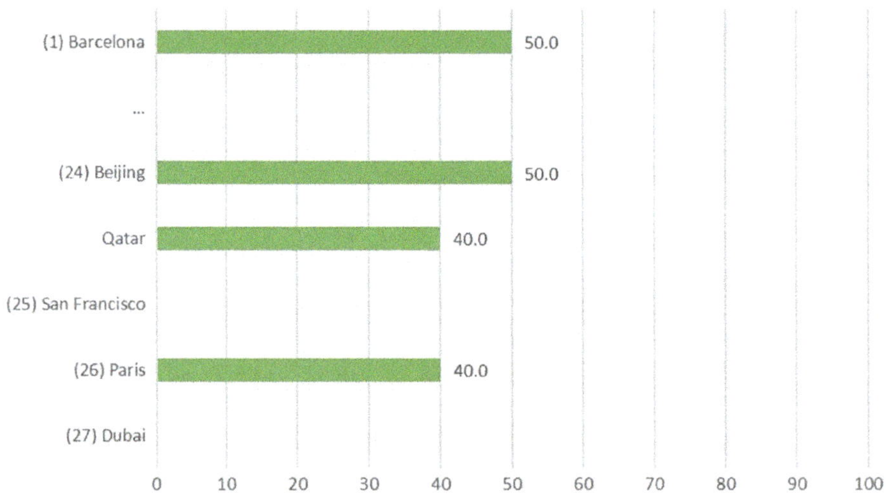

Fig. 7.6 Pillar five: Participation. Compared to other cities Qatar reaches a quite good result, but overall it is rather just acceptable outcome. *Source of foreign data* Fietkiewicz et al., (2017)

intuitively is plausible for the users, but in this case not given. The circumstance that many advertised services do not work in the end is also unfavorable. Links to most external webpages often do not work neither.

7.2 E-Government Social Media Study

In the beginning of our e-Government social media (or Government 2.0) study we retrieved a list from Qatar's e-Government website containing 35 governmental institutions and checked, whether those institutions are active on both social media platforms, Facebook and Instagram, or just on one of them, and examined the functionality and availability of the given links. After careful consideration we ascertained six institutions sharing a link on their website that is not available or functional: Qatar Financial Markets Authority, Communications Regulatory Authority, Ministry of Foreign Affairs, Ministry of Finance, Ministry of Energy and Industry and Ministry of Economy and Commerce. The Ministry of Environment has stated an inactive Facebook link, but is still a part of our list (Table 7.1) because the Instagram account is active. The same holds true for some institutions with inactive or nonexistent Instagram accounts and active Facebook accounts: Civil Aviation Authority, Ministry of Administrative Development, Ministry of Culture, Arts and Heritage, Ministry of Development Planning and Statistics, Ministry of Information and Communications Technology, Ministry of Municipality and Urban Planning, Mowasalat, Qatar Assistive Technology Center, Qatar Museums Authority, Qatar Postal Services Company—Qpost, Sheikh Tamim bin Hamad bin Khalifa Al Thani and Supreme Council of Health. Since the absence of data does not create added value, we excluded those institutions from our following analysis. Hence, the list was shortened to 29 institutions.

Overall, 28 of the institutions have an active account on Facebook. There are only 17 institutions with active accounts on Instagram. Just 16 institutions are active on both platforms, which is a little more than half of all examined institutions.

7.2.1 Data Acquisition

Before we start describing the findings, we outline our data acquisition of the 45 webpages in detail. We will start the explanation with the social media platform Facebook. On every Facebook page, we can find the number of followers of the page that is mandatorily needed to evaluate the popularity of a social media page. Also on Instagram the number of followers is easy to find on the respective profile. To find out the exact number of likes, comments, shares as well as the language the Facebook post is written in, we analyzed every post separately.

Table 7.1 Institutions and their availability on Facebook and Instagram

Institution	Facebook	Instagram
Childhood culture center	Yes	Yes
Civil aviation authority	Yes	No
General authority of customs	Yes	Yes
Hamad medical corporation	Yes	Yes
Hukoomi—Qatar e-Government	Yes	Yes
Ministry of administrative development	Yes	No
Ministry of culture, arts and heritage	Yes	No
Ministry of development planning and statistics	Yes	No
Ministry of endowments and islamic affairs	Yes	Yes
Ministry of environment	No	Yes
Ministry of ICT	Yes	No
Ministry of interior	Yes	Yes
Ministry of justice	Yes	Yes
Ministry of labor and social affairs	Yes	Yes
Ministry of municipality and urban planning	Yes	No
Ministry of transport	Yes	Yes
Ministry of youth and sports	Yes	Yes
Mowasalat	Yes	No
Public works authority	Yes	Yes
Qatar assistive technology center	Yes	No
Qatar chamber	Yes	Yes
Qatar general electricity and water corporation	Yes	Yes
Qatar museums authority	Yes	No
Qatar postal services company—QPOST	Yes	No
Qatar tourism authority	Yes	Yes
Regulatory authority for charitable activities	Yes	Yes
The qatari emir sheikh tamim bin hamad bin khalifa al thani	Yes	No
Supreme council of education	Yes	Yes
Supreme council of health	Yes	No

During the study period from January 2nd, 2016 to January 15th, 2016 we went online every day to reassess whether there was a new post or not. Then, the number of likes, which is important for evaluating the approval of a post, was listed. As an indicator of the actual importance of a post we also noted the number of shares. Afterwards, the number of comments was specified to indicate the commitment for a fluctuating content exchange between the citizens and the government (Bonsón & Ratkai, 2013). Usually, if the post was not just an image, we additionally noted down the language of the post—if it was in Arabic, English, or both—to be able to make a statement about the target groups of the posts in the end of our evaluation. Due to the fact that many expats are living in Doha and the surrounding region, it is not unimportant to write multilingual posts. If they are written only in Arabic, many expatriate followers and readers might miss the point of the content. A special case of a post is an image without any text. Here, we listed Arabic and English as the languages used, because we assume that everyone can understand the content of an image—no matter which language he or she may speak. The same indicators were applied to evaluate Instagram posts.

7.2.2 Followers

After reworking the list of active Facebook and Instagram accounts, we first looked up the actual number of followers on Facebook (Fig. 7.7) and on Instagram

Fig. 7.7 Number of Facebook followers (per institution). *Source* Facebook, June 8, 2016

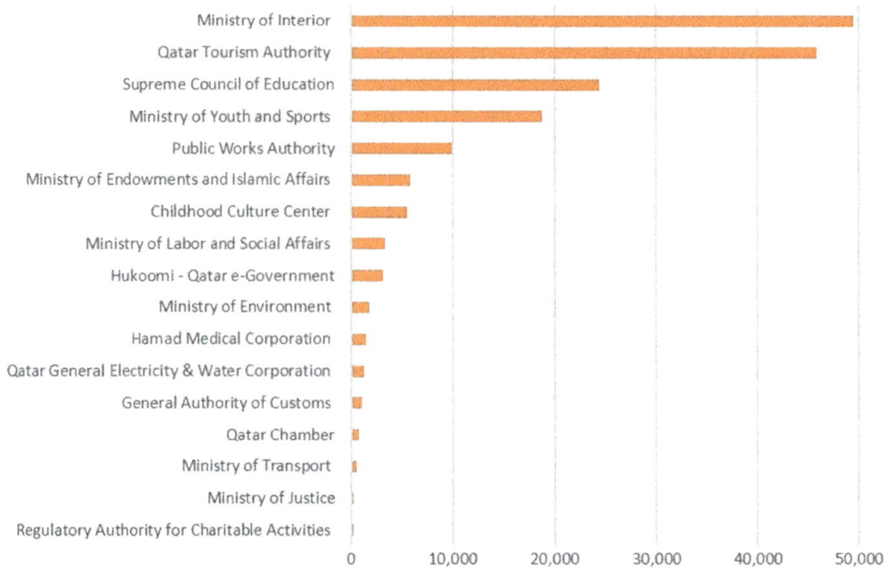

Fig. 7.8 Number of Instagram followers (per institution). *Source* Instagram, June 8, 2016

(Fig. 7.8) to bring the institutions' general popularity into quest. If analyzed per institution, all of the following findings are ordered descending by frequency. Furthermore, to better distinguish the social media channels from each other, the following graphs show Facebook data in blue (dark grey), and Instagram data in orange color (light grey).

Front-runner regarding the number of Facebook followers is the Qatar Tourism Authority with overall just under 500,000 followers. The second highest number of followers is reached by the Ministry of Interior, but the number is just a little higher than half of that of the Tourism Authority. The Ministry of Endowments and Islamic Affairs and the Qatar Museums Authority have high follower numbers as well. With the Supreme Council of Health the number of followers skids under the 100,000 level right up to under 10,000 followers for the Ministry of Justice.

Concerning the Instagram results, the maximum of followers is in comparison strikingly lower. With just 49,000-odd followers the Ministry of Interior reaches the first rank in this category. The Tourism Authority has nearly the same number of followers so that we can ascertain that the top performers on Instagram are the same as on Facebook—just in different order. Two other institutions reach follower numbers of around 20,000: the Supreme Council of Education and the Ministry of Youth. The number of followers of the other institutions is evidently lower, starting with values under 10,000 and decreasing to values of a few hundred in total.

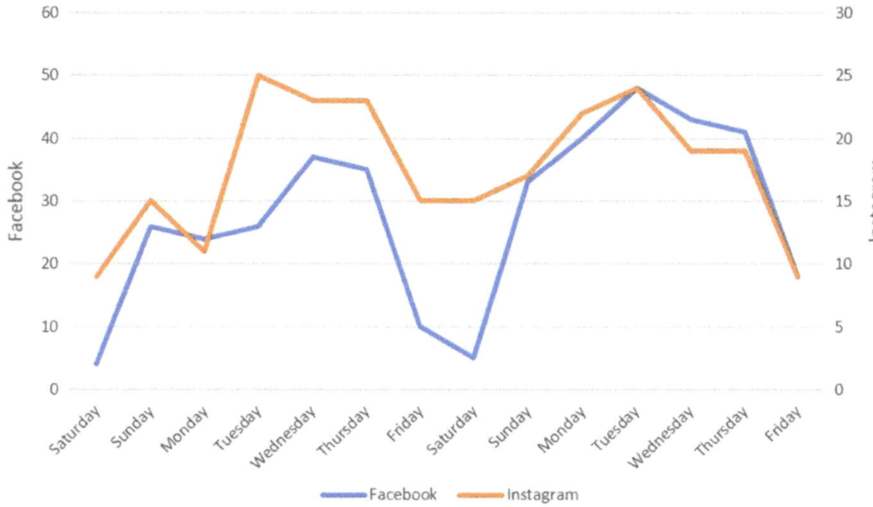

Fig. 7.9 Number of posts on Facebook and Instagram (per day)

7.2.3 Posts

In a next step, we sum up the posts per day of all the governmental bodies, once for Facebook and once for Instagram, to see on which days the institutions are most active on their social media sites (Fig. 7.9).

Taken as a whole, we can say that the activity starts at the beginning of the workweek (Sunday), increases during the week and decreases with the beginning of the weekend (Friday). The curves—of Facebook and Instagram—run rather parallel to each other. Relative to the estimated service life, the curves will probably be in antiphase with the citizens' social media use because most of the people will have most time on the weekend to like, comment and share posts or images. This might be due to the fact that most of the government employees work on weekdays, but it is in all likelihood unprofitable for collecting reactions and certainly also new followers.

The number of Facebook and Instagram posts is an indicator on the overall social media activity of—in our case—different governmental bodies. To learn more about the distribution of those posts Fig. 7.10 can be used to analyze the Facebook results.

Ordered by the number of posts, the Ministry of Endowments and Islamic Affairs does best in this category (3.86 posts on average per day). With an average of 3.50 posts, the Civil Aviation Authority posted nearly the equal number. At this point, we can already determine that the institution with the highest number of followers is not the institution with the most posts as a matter of routine. So just because the interest in the institution is on a high level, the demand is not quite satisfied. The other way around, the general assumption that "more activity engages

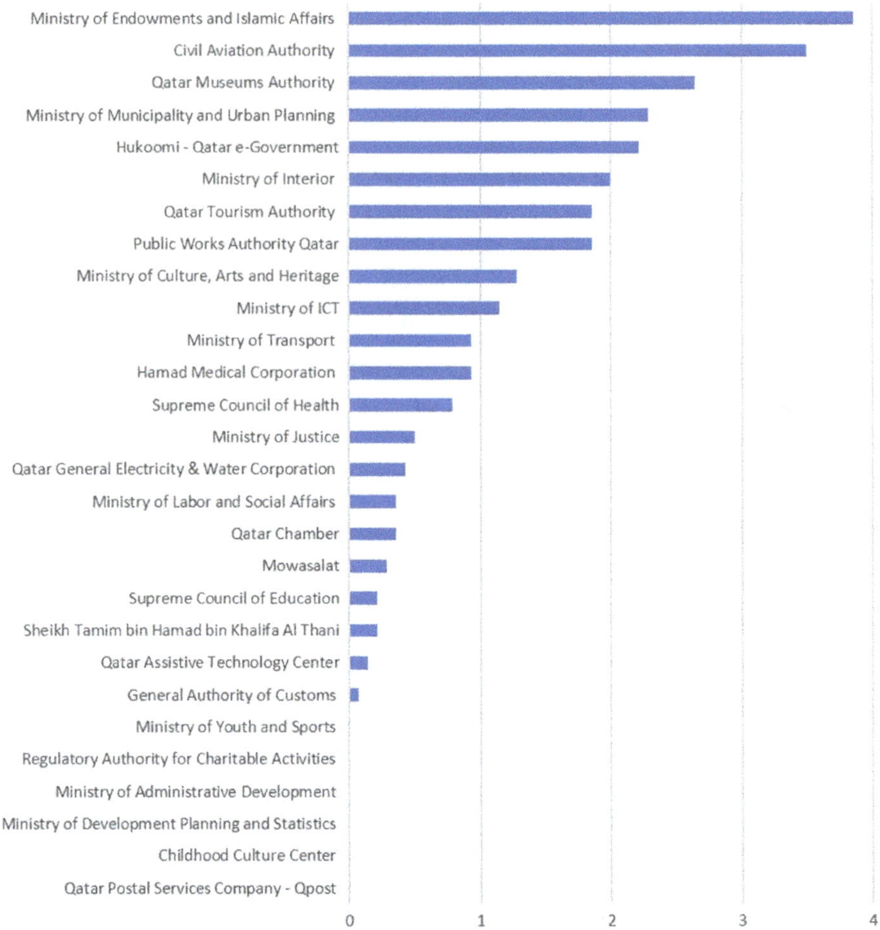

Fig. 7.10 Average number of Facebook posts per day (per institution)

more users" (Mainka et al., 2015, p. 250) does not seem to be right in this case. Instead, the institutions get their high number of followers probably not by posting as much as they can, but rather by posting interesting content. The number of posts decreases fast to an average of under 3.00 posts per day regarding the remaining government bodies and ends with six institutions that did not post anything during the study time, starting with the Ministry of Youth and Sports.

As we are interested in both channels, we also counted the Instagram posts per institution (Fig. 7.11). In this category, the Ministry of Youth and Sports publishes the most with average 5.07 posts per day. Relative to the number of Facebook posts, this is rather surprising, because the number of more elaborately designed Instagram posts, where always a photo must be included, is decidedly higher. "The posting frequency should correspond with the type of service" (Mainka et al., 2015,

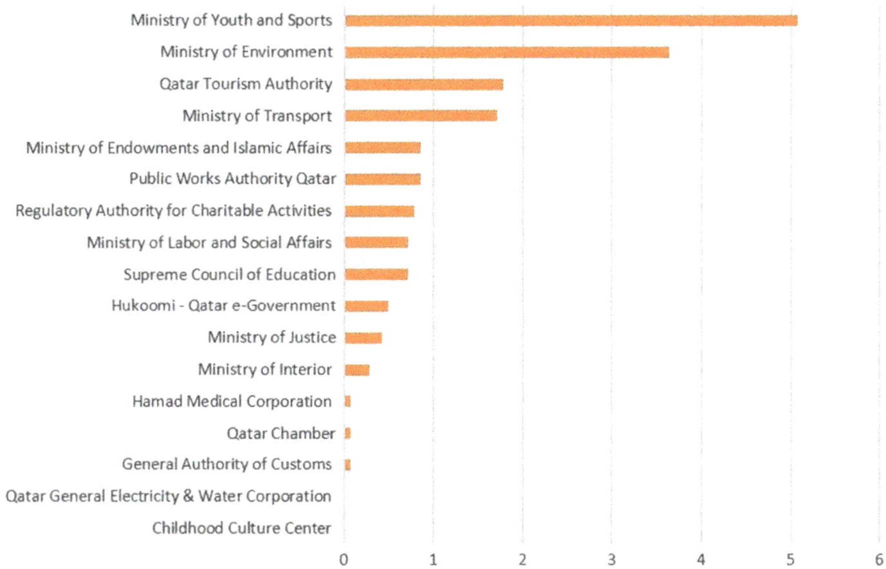

Fig. 7.11 Average number of Instagram posts per day (per institution)

p. 250). In this case, the posting frequency is higher than expected. The Ministry of Environment published a little less, the Qatari Tourism Authority and the Ministry of Transport just half as much. With the Ministry of Endowments and Islamic Affairs and the Public Works Authority, the number drops steadily under an average of 0.86 posts per day and ends up with two institutions that did not post anything during the study time, the Childhood Culture Center and the Qatar General Electricity and Water Corporation. In total, twenty-two Facebook and fifteen Instagram sites were analyzed and are presented in the following findings.

7.2.4 Post Analysis: Likes, Comments, Shares and Language

We start our Facebook post analysis with the evaluation of the average number of likes—the probably most significant factor for determining the popularity and approval of a post (Fig. 7.12).

In this category, an institution or more precisely a person comes to the fore that did not stand out, yet—the Emir. He receives 2804.67 likes on average per post. In comparison to the analysis above, this is surprising because the numbers of followers and posts are rather low in contrast to other authorities. With less than one-third of the Emir's average likes per post, the Qatari Tourism Authority (895.27 likes on average) and the Ministry of Interior (807.39 likes on average) are placed

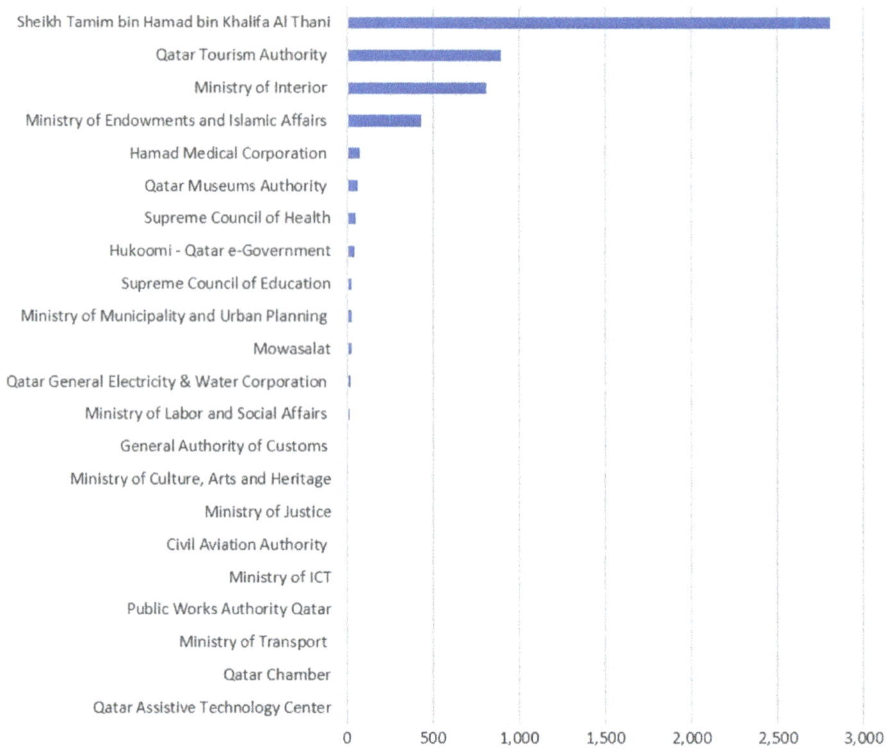

Fig. 7.12 Average number of likes per post on Facebook (per institution)

second and third. Here, the high numbers of likes per post come along with high numbers of followers and posts. The same holds true for the Ministry of Endowments and Islamic Affairs that has, on average, half the number of likes, but many followers and posts. The number decreases significantly when looking at the following institutions: Hamad Medical Corporation reaches on average 76.54 likes and Qatar's Museums Authority 62.97 likes. In comparison to this, the likes of all the other institutions' posts decrease on and on and are, for some institutions with average values between 49.55 and 0, rather negligible.

Significantly less comments—that are an indicator of active user behavior and in governmental matters maybe attempts of e-Participation—were posted. The highest value is reached by the Emir with an average of 190.00 (Fig. 7.13).

In comparison to the other objects of study, the number of comments is in general much higher. The first place in our ranking corresponds, therefore, with the number of likes and is surprising in view of the comparably low numbers of followers and posts. The reactions to the Emir's posts are plainly visible, a communication is possible and a government-citizen relationship might occur. Some kind of e-Participation, a part of e-Government where the "citizens are provided, and seize, the opportunity of engaging in political decision-making processes"

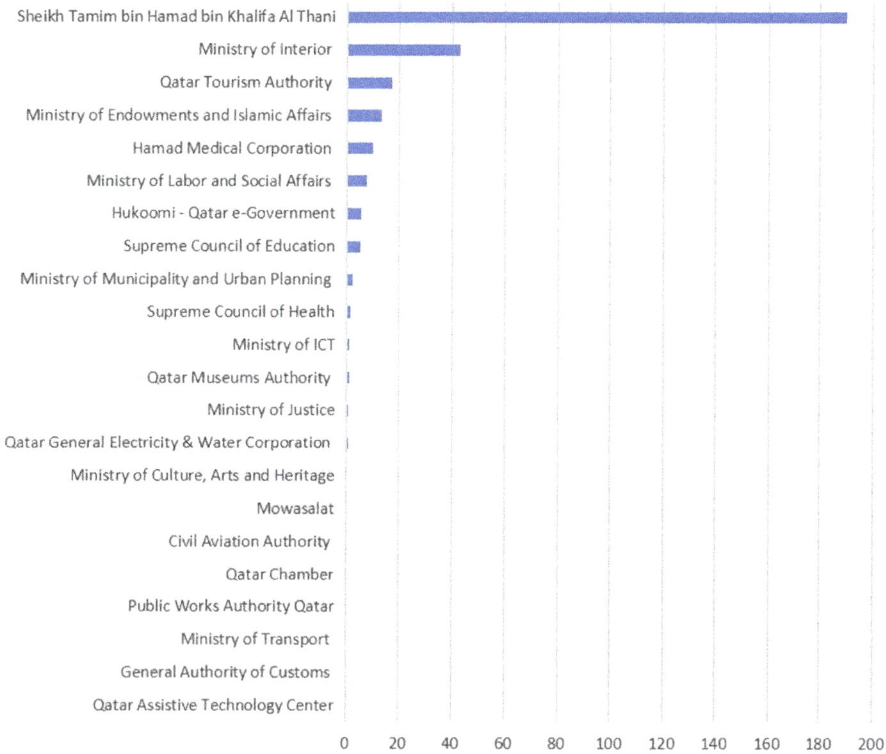

Fig. 7.13 Average number of comments per post on Facebook (per institution)

(Mainka, Hartmann, Stock, & Peters, 2014, p. 1715), becomes visible. With less than a quarter of comments (43.07), the Ministry of Interior is ranked second. Three institutions reach an average value between 10.00 and 20.00 comments per post: Qatar Tourism Authority (17.08), Ministry of Endowments and Islamic Affairs (13.06) and Hamad Medical Corporation (10.23). The Ministry of Labor and Social Affairs comes closest to the top five with an average of 7.60 comments per post. All the other governmental bodies have on average under 6.00 comments per post (some even 0.00). In these cases, we can no longer speak about a representative e-Participation.

In a next step regarding the Facebook posts, we analyzed the average number of shares, where shares are an indicator on spreading recent relevant content. The distribution is similar to the distribution of the comments. Front-runner is once again the Emir with an average of 191.00 shares per post (Fig. 7.14).

With a little more than a third of this number (68.04 shares on average per post) and a little less than a third (61.31 shares on average per post), the Ministry of Interior is ranked second and the Qatar Tourism Authority third. Considering the topicality of a post, it is possible to say that the three before mentioned authorities post the most topical content regarding the current situation. The Ministry of

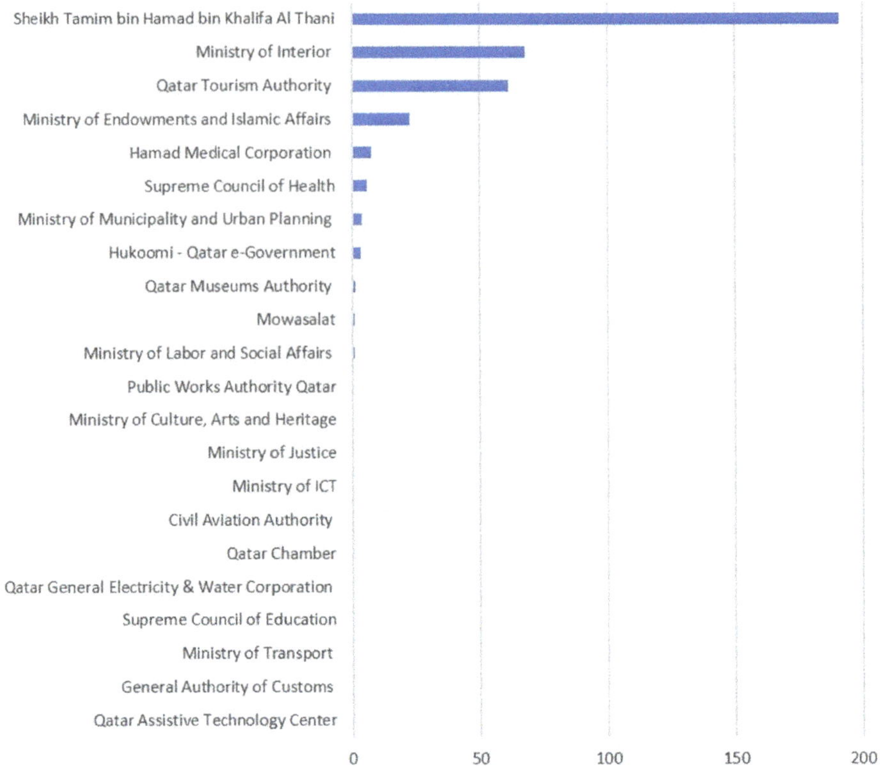

Fig. 7.14 Average number of shares per post on Facebook (per institution)

Endowments and Islamic Affairs is doing fairly well with an average of 22.61 shares per post. Two institutions range between an average of 6.00 and 8.00 shares per post—the Hamad Medical Corporation and the Supreme Council of Health—and are, therefore, the last two noteworthy institutions regarding the number of shares on Facebook. With values between almost 4.00 and 0.00 the remaining institutions can be considered irrelevant for the analysis and the statement about the social media popularity of the Qatari e-Government when it comes to the number of shares.

To ascertain the most accepted and supported government body on Instagram, the same evaluation is made for this social media platform. Figure 7.15 shows the rank order relating to the number of likes—that measures the approval of a post—of the 15 institutions that uploaded an image during the period of study.

Considering these results, the Ministry of Interior is the conspicuously most minded institution (617.25 likes on average per post). The extent of Instagram likes is different from the one on Facebook—up to 617.25 likes on average per image on Instagram versus an average of up to 2804.67 likes per post on Facebook. With about one-third of the likes of the Ministry of Interior, the Tourism Authority is ranked second. The Supreme Council of Education achieves the third place in this

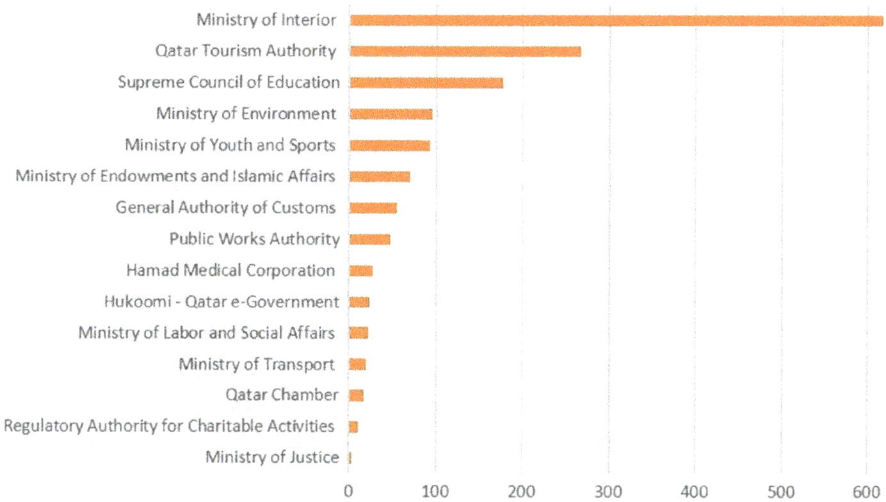

Fig. 7.15 Average number of likes per image on Instagram (per institution)

category. The Ministry of Environment and the Ministry of Youth and Sports have quite a lot likes as well. Regarding the average number of likes, the values of the remaining institutions are between a relatively high number of 70.58 for the Ministry of Endowments and Islamic Affairs and a very low number of 3.00 for the Ministry of Justice.

Relating to the comments measuring the government-citizen interaction, the Supreme Council of Education does best in this category with an average value of 16.40 (Fig. 7.16).

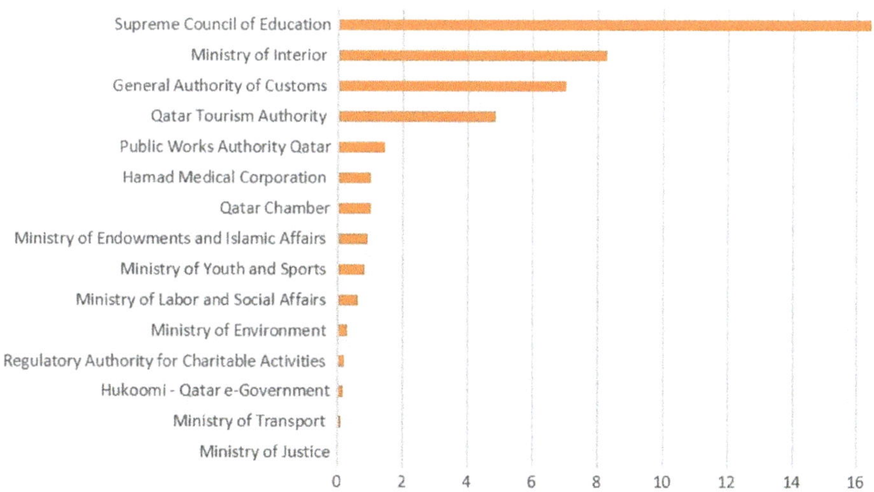

Fig. 7.16 Average number of comments per image on Instagram (per institution)

With on average the half of the number of comments the Ministry of Interior is ranked second (on average 8.00 comments per image). Close to this value, the General Authority of Customs is ranked one place behind. Four institutions range between almost 5.00 and 1.00 average comments per post: Qatar Tourism Authority, Public Works Authority, Hamad Medical Corporation and Qatar Chamber. The average number of comments per image for these institutions is insignificant. It is clearly recognizable that the values start at a lower level than those of Facebook, so that it is hard to say whether a government-citizen communication indeed exists, but as a result of the comparison with the Emir's Facebook feedback it seems to be limited.

Due to the high number of expats who probably do not all speak Arabic fluently, we try to ascertain whether all citizens have access to the information posted on the social media channels and, therefore, can participate in the government-citizen communication. On that account, for each institution we counted the number of posts on Facebook that are written in the most spoken language, English, as well as in the native language Arabic. Of course, only the institutions that posted anything during the two-week period are considered. The results of the summation that are first ordered by the number of institutions posting in both languages, second by those posting just in English and third by those writing just in Arabic are concluded in Figs. 7.17 and 7.18.

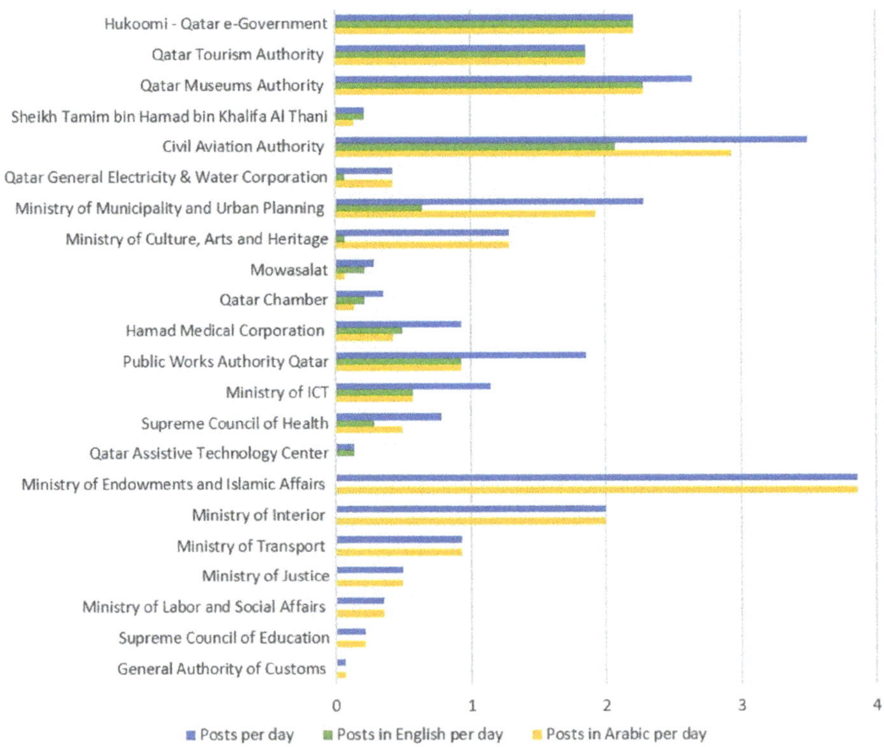

Fig. 7.17 Language of posts on Facebook (per institution)

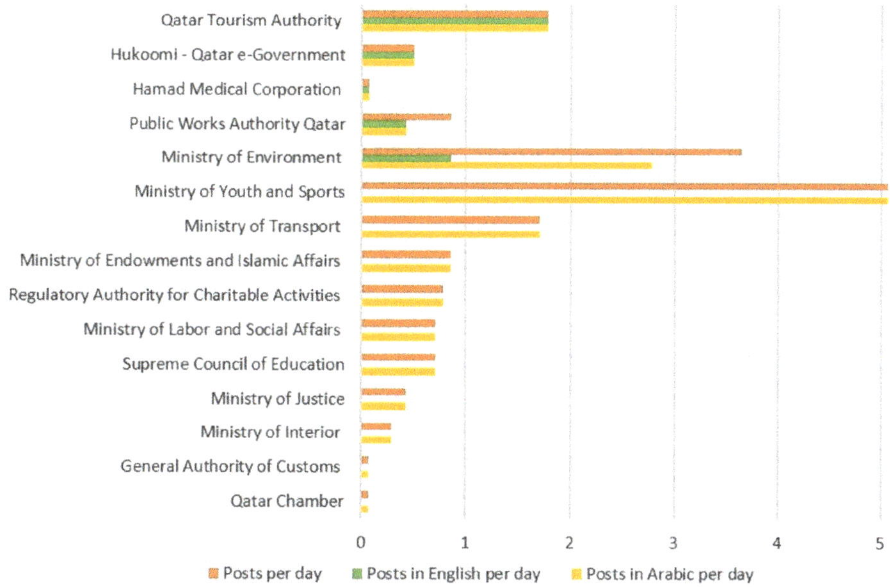

Fig. 7.18 Language of posts on Instagram (per institution)

Not more than two of the overall twenty-two institutions that post something on Facebook write their texts in English and Arabic (Hukoomi and Qatar Tourism Authority). Every text is translated from Arabic into English or vice versa so that they can be published in just one post. Qatar Museums Authority did not post every text in both languages, but overall 86.49% of the posts are in English and 86.49% of the posts are in Arabic. The Emir of Qatar posts rather in English (100.00%) than Arabic (66.67%). This is exactly the other way around when regarding the Civil Aviation Authority (English: 59.18%; Arabic: 83.67%) and the Qatar General Electricity and Water Corporation (English: 16.67%; Arabic: 100.00%). There are eight other institutions that post in both languages—some more in English, some more in Arabic. Just one institution posts without exception in English (Qatar Assistive Technology Center) and the other seven exclusively in Qatar's native language Arabic. Overall, most of the institutions post in a way every inhabitant can understand.

To get the respective results for the Instagram posts, we also counted the English and the Arabic descriptions of the images for each institution (Fig. 7.18).

On Instagram, we have three institutions posting every text in both languages: Qatar Tourism Authority, Hukoomi (Qatar's official e-Government online site) and Hamad Medical Corporation. The Public Works Authority posted 50.00% in English and 50.00% in Arabic, where it must be assumed that every English post has a corresponding Arabic post and vice versa. Unlike the previously mentioned approach, this institution published a new post for each translation. The Ministry of Environment writes in the two languages as well, but in most cases the language is

Arabic (76.47% Arabic versus 24.53% English). All the other analyzed institutions post only in the native language Arabic.

7.2.5 *Government 2.0: Ranking of Institutions*

Table 7.2 summarizes the results of the previous analysis and outlines the top 10 institutions in the following categories: followers on Facebook and Instagram, posts on Facebook and Instagram, average number of likes, comments and shares on Facebook as well as average number of likes and comments on Instagram.

Considering the overall average, Qatar Tourism Authority is on the top and, therefore, the most popular, most accepted and most communicative e-Government body in Qatar, within the chosen institutions. Indeed, the authority did not post the most content, but the reactions and the general public interest in the institution and its posts is very high as can be noticed by looking at the followers, likes, shares and comments values. The fact that all posts are in English and Arabic certainly contributes to this success, because this way every citizen, regardless of whether he is Qatari or expat, understands the information and is able as well as more willing to react to it. The Tourism Authority—or better known as VisitQatar on the social media channels—posts almost always about culture, sports and shopping, provides different leisure-time opportunities to Qatar's inhabitants and visitors and involves the users with incentive questions (Fig. 7.19).

In addition to it, it often refers to internal events, like Qatar Motor Show, different soccer games or concerts. It presents the city, its landscape and leisure facilities in the proper light and underlines the posted content also on Facebook always with an image or video and, hence, attracts the public attention. Sometimes, it does advertising for the city's or its own account.

The Ministry of Interior is ranked second. Regarding the number of followers, this ministry and the Tourism Authority take the two best places—Qatar Tourism Authority does best in terms of Facebook and second in terms of Instagram, and the Ministry of Interior does best in the number of Instagram followers and second in the category followers on Facebook. The ministry did not post the most content of all institutions and once lands on the 12th rank (average number of Instagram posts), but it gets one of the best feedbacks—in terms of likes on Instagram even the best one. As the name suggests, the topics of the ministry's posts revolve around domestic policy. Aspects like security, criminality and even infrastructure and traffic are subjects of the posting. All of which have an effect on daily routines. As a result of the continuous usage of text-image-video-combinations on both channels, the response is great. However, due to the fact that all postings are just in Arabic, a language that not all inhabitants understand, it is probably lower than it could be. With the help of an English translation, the feedback could be even larger.

The Ministry of Endowments and Islamic Affairs is ranked on the first place regarding the number of posts on Facebook, which admittedly indicates a very active institution, but not necessarily a motivated government-citizen relationship

Table 7.2 Evaluation of the different indicators followers, posts, likes, comments and shares on Facebook and Instagram

Rank	Follower on facebook (Number)	Follower on instagram (Number)	Posts on facebook per day (Average number)	Posts on instagram per day (Average number)	Likes on facebook per post (Average number)	Comments on facebook per post (Average number)	Shares on facebook per post (Average number)	Likes on instagram per post (Average number)	Comments on instagram per post (Average number)	Overall (Average rank)
1.	Qatar Tourism Authority (499,631)	Ministry of Interior (49,400)	Ministry of Endowments and Islamic Affairs (3.86)	Ministry of Youth and Sports (5.07)	Sheikh Tamim bin Hamad bin Khalifa Al Thani (2804.67)	Sheikh Tamim bin Hamad bin Khalifa Al Thani (190.00)	Sheikh Tamim bin Hamad bin Khalifa Al Thani (191.00)	Ministry of Interior (617.25)	Supreme Council of Education (16.40)	Qatar Tourism Authority (3.00)
2.	Ministry of Interior (270,052)	Qatar Tourism Authority (45,800)	Civil Aviation Authority (3.50)	Ministry of Environment (3.64)	Qatar Tourism Authority (895.27)	Ministry of Interior (43.07)	Ministry of Interior (68.04)	Qatar Tourism Authority (267.40)	Ministry of Interior (8.25)	Ministry of Interior (3.44)
3.	Ministry of Endowments and Islamic Affairs (254,494)	Supreme Council of Education (24,400)	Qatar Museums Authority (2.64)	Qatar Tourism Authority (1.79)	Ministry of Interior (807.39)	Qatar Tourism Authority (17.08)	Qatar Tourism Authority (61.31)	Supreme Council of Education (177.50)	General Authority of Customs (7.00)	Ministry of Endowments and Islamic Affairs (4.56)
4.	Qatar Museums Authority (208,239)	Ministry of Youth and Sports (18,800)	Ministry of Municipality and Urban Planning (2.29)	Ministry of Transport (1.71)	Ministry of Endowments and Islamic Affairs (426.61)	Ministry of Endowments and Islamic Affairs (13.06)	Ministry of Endowments and Islamic Affairs (22.61)	Ministry of Environment (96.06)	Qatar Tourism Authority (4.84)	Sheikh Tamim bin Hamad bin Khalifa Al Thani (6.60)
5.	Supreme Council of Health (95,344)	Public Works Authority (9980)	Hukoomi—Qatar e-Government (2.21)	Public Works Authority (0.86) Ministry of Endowments and Islamic Affairs (0.86)	Hamad Medical Corporation (76.54)	Hamad Medical Corporation (10.23)	Hamad Medical Corporation (7.62)	Ministry of Youth and Sports (92.55)	Public Works Authority (1.42)	Ministry of Environment (6.75)
6.	Childhood Culture Center (57,943)	Ministry of Endowments and Islamic Affairs (5849)	Ministry of Interior (2.00)		Qatar Museums Authority (62.97)	Ministry of Labor and Social Affairs (7.60)	Supreme Council of Health (6.09)	Ministry of Endowments and Islamic Affairs (70.58)	Hamad Medical Corporation (1.00) Qatar Chamber (1.00)	Qatar Museums Authority (6.80)

(continued)

Table 7.2 (continued)

Rank	Follower on facebook (Number)	Follower on instagram (Number)	Posts on facebook per day (Average number)	Posts on instagram per day (Average number)	Likes on facebook per post (Average number)	Comments on facebook per post (Average number)	Shares on facebook per post (Average number)	Likes on instagram per post (Average number)	Comments on instagram per post (Average number)	Overall (Average rank)
7.	Hukoomi—Qatar e-Government (57,821)	Childhood Culture Center (5533)	Public Works Authority (1.86) Qatar Tourism Authority (1.86)	Regulatory Authority for Charitable Activities (0.79)	Supreme Council of Health (49.55)	Hukoomi—Qatar e-Government (5.81)	Ministry of Municipality and Urban Planning (4.06)	General Authority of Customs (55.00)		Hamad Medical Corporation (8.11)
8.	Hamad Medical Corporation (53,706)	Ministry of Labor and Social Affairs (3359)		Ministry of Labor and Social Affairs (0.71) Supreme Council of Education (0.71)	Hukoomi—Qatar e-Government (44.29)	Supreme Council of Education (5.33)	Hukoomi—Qatar e-Government (3.35)	Public Works Authority (47.67)	Ministry of Endowments and Islamic Affairs (0.92)	Supreme Council of Health (8.20)
9.	Ministry of ICT (44,634)	Hukoomi—Qatar e-Government (3172)	Ministry of Culture, Arts and Heritage (1.29)		Supreme Council of Education (28.33)	Ministry of Municipality and Urban Planning (2.38)	Qatar Museums Authority (1.59)	Hamad Medical Corporation (27.00)	Ministry of Youth and Sports (0.80)	Hukoomi—Qatar e-Government (8.56)
10.	Public Works Authority (40,208)	Ministry and Environment (1805)	Ministry of ICT (1.14)	Hukoomi—Qatar e-Government (0.50)	Ministry of Municipality and Urban Planning (26.91)	Supreme Council of Health (1.55)	Mowasalat (1.00)	Hukoomi—Qatar e-Government (22.86)	Ministry of Labor and Social Affairs (0.60)	Ministry of Municipality and Urban Planning (8.80)

Fig. 7.19 Facebook posts of Qatar's e-Government top performer in social media usage: Tourism Authority. *Source* Facebook, July 11, 2016

especially as the Facebook likes, comments and shares values could be higher. A high average number of likes, shares and comments could have indicated the users' interest or feedback and lead to a communication between both parties. Because of this shortcoming, the Ministry of Endowments achieves an overall rank of 3. The postings mostly deal with cultural and religious topics. The Ministry of Endowments and Islamic Affairs' intention is to bring the Arabic culture and way of life close to the users. This, however, happens in Arabic language and most times with a support of short texts on Instagram and in various combinations on Facebook. The population seems to be just partly keen on it. Indeed, the number of followers is high, but reactions to the posts are somewhat low in contrast to the better-placed institutions (about half as much). Despite the rather not that good assessment the ministry achieves rank 8 among the top 10 placements.

It looks entirely different when regarding the Emir's results. Here, everything is about the likes, comments and shares of the posts. In this case, we speak about performance on Facebook, because he is active only on this channel. In all the three categories, he achieves the first rank and, despite the fact that he does not post very much, his postings gain much recognition. He limits his postings to images most of the times and just in some cases adds texts that turn out rather poorly—as for example phrases like "Like/Share"—so that no one faces a challenge when trying to follow his news. He posts about both, his private and professional life as a model for all Qataris and the nation. Especially due to a rather low number of followers and a rather low number of posts, the frequent, continuous feedback and communication with the citizens is particularly noteworthy. Owing to two other nominations inferior to the top 10 and no scores in the Instagram categories, the Emir's account finished up in an overall rank of 4.

The fifth most popular institution is the Ministry of Environment. This institution is only active on Instagram, but its posted information goes down well with the public which is shown due to the mere fact that the average number of likes per picture is high. The result is quite good, which in turn may be attributed to the multilingual positing behavior. The posts are mainly about general environment, environmentalism or environmental policy. The ministry makes an effort to turn the spotlight on this topic by posting a high number of images described by a short text. They enjoy popularity, which can be recognized by the relatively high number of likes per post. The translation of every post into English could help in circulating this important topic.

Qatar Museums Authority is ranked sixth in the overall analysis. Here, the focus moves to a relatively high number of followers and posts. This institution is likewise represented on just one social media channel, namely Facebook, and when compared to others, does not earn many likes for its postings. In terms of content, this institution concentrates almost entirely on cultural topics like architecture, artists, collections and certainly museums. In addition to it, there are some leisure advices and a few postings concerning knowledge topics—but in every case the posts are related to culture or arts. Most of the posts are written in Arabic and English so that they are easy to understand—also for the expats. However, even if the content is similar to the one of Qatar Tourism, this account is more poorly accepted by the population, which might be not due to the frequency, language or design of the posts, but rather due to their content.

The Hamad Medical Corporation, placed on an overall rank of 7, pops up mostly on relatively good places in the Facebook categories. In terms of likes, comments and shares, the institution achieves each time the 5th place, which indicates a moderate content and government-citizen relationship. In relation to the rather low number of followers, the proportion of the results of likes, comments and shares analyses are comparable to those of the Ministry of Endowments and Islamic Affairs that is ranked third. The content posted by the Hamad Medical Center is entirely about health, which seems to be a topic in which the citizens have interest. In most cases, the postings consist of texts and images. Sometimes links to other medical institutions are added. Since most of the posts are translated from Arabic into English, everyone is able to participate in discussions on this important subject.

The results of the Supreme Council of Health look similar to some of the accounts mentioned above. This institution is only active on Facebook. Here, no better placement than rank 5 (Facebook followers) is achieved. However, on average the ranking is poorer so that only the 8th rank could be achieved. Regarding likes, comments and shares, the council performs relatively weak. With regard to the content, we can summarize the topics: healthcare, healthcare policy and news about healthcare information. In most cases, they refer to magazines, other information or their own website, and design their postings clearly with images and texts. Taken as a whole, the Supreme Council of Health performs poorer than the competitive Hamad Medical Corporation.

The official Qatari e-Government platform Hukoomi is ranked on the 9th place. Since Hukoomi achieves in every case a 5th rank or worse, no better placement was possible. The others have more likes, comments and shares. Hukoomi does not have a very high number of followers, but publishes a relatively high number of posts—almost entirely a combination of texts, images, videos or links—but they get barely any feedback, at least on Instagram. And yet, they offer a full range of topics like healthcare, ICT, e-Government, traffic and general greeting to get in contact with the citizens. It is possible that the selection is not adequate or too various for the citizens' tastes. However, as a universal institution it is probably hard to commit to one specific topic. For sure the language is not the barrier, because both, the Arabic and the English coverage, amount to 100%.

The Ministry of Municipality and Urban Planning—ranked on the 10th place of our ranking—is active only on Facebook. The ministry achieves top 10 ranks in the categories posts, likes, comments and shares on Facebook. Topics that are discussed are wide-ranging: from health to environment and the weather. The postings are sometimes in Arabic, sometimes in English and sometimes in both languages.

Overall, we can say that the posting frequency as well as the acceptance, interaction and topicality do not entirely correlate. It goes without saying that the number of followers has a wider influence on those factors: the more followers, the more readers and in turn the higher chance of finding people that are interested in the content. Then again, without interesting content it is more difficult to recruit new followers. Thus, the governmental institutions have to find out what the citizens need and have to adapt their content to the citizens' wishes. However, despite such attempts, some topics will always be more interesting than others and due to the fact that every institution has to remain true to itself, it would be hard for some of them to become more successful. A few of the institutions have some room for improvement and have to find out what the citizens miss—perhaps with the help of surveys or other suitable methods. Some of the "citizens are still not fully aware of e-government services, are concerned about security and some are hindered by the lack of internet access" (Al-Shafi & Weerakkody, 2010, p. 19). Do these problems rub off on the social media activity of the government and the social media use of the citizens? Only if both parties are willing to overcome the difficulties, it is possible to develop a government-citizen relationship with the help of e-Participation, which until now was achieved by a rather small number of administrative bodies. This way they can give high confidence to the inhabitants and will get satisfied customers, feeling well and understood in their own country.

7.2.6 Government 2.0 and E-Government Services

The study shows that some institutions are fully engaged and interested in informing the citizens and in exchanging information with them. The implementation of their e-Government strategies seems to be successful and sufficient, so that the response rate to their posts is high and a citizen-government exchange visible. However, others fail and neglect the task of being responsive to the citizens via social media. Especially, when regarding Hukoomi Qatar the results are rather surprising. As it is the official e-Government platform, we assumed a better ranking than 9. The results of the study confirm the overall meaning that e-Government in Qatar is not more than mediocre. Perhaps, these findings result from the fact that the approach of implementation proceeds in a rather unmethodical way (Interview partners 4 & 9). In general, Qatar's e-Government (including different e-Government services like the payment of tickets or the like) "has not achieved the necessary level of success in terms of implementation and diffusion" (Weerakkody, El-Haddadeh, & Al-Shafi, 2011, p. 173). Indeed, all GCC countries have improved their ICT use and governmental online services and have promoted wider access to the internet, but overall e-services are not entirely satisfactory these days (Al-Shafi & Weerakkody, 2009; Mansour, 2012). The meanings of the citizens' contribution to governmental decision-making process—one of the most important functions of e-Government—are different: On the one hand interview partner 1 said that "a decision maker makes a decision and everyone follows his actions"—regardless of whether it makes sense or not. In addition to it, the inhabitants are only rarely integrated in the government's intentions, no one knows what the others do and they do not make any efforts to change this (Interview partners 4 & 6). On the other hand, others told us that it is just bureaucracy and an exchange exists despite all that (Interview partner 5). Citizens "can make things happen, have influence and can voice themselves" (Interview partner 12). These clearly different statements show that e-Government does not go down well with the overall public and that it has not made a sufficient effort until now. With those e-Services that are available, the customers are nevertheless contented (Al-Yafi, Osman, & Hindi, 2014) and also social media sites that are well-kept, like those of Qatar Tourism or the Ministry of Interior, seem to satisfy the users' needs.

7.3 Summarizing Qatar's E-Government Strategies

To sum up, Qatar and its government institutions seem to be on the right track of implementing e-Government services including the interaction via social media channels, but until now, only few services and social media sites are implemented successfully. Just some of the institutions' social media sites are today really popular in the country and interesting for the citizens. However, most of them have to refine themselves to become successful providers and conversationalists. For some of them, it could—in terms of social media—become a big challenge due to the fact that some topics like leisure, sports and shopping are simply more interesting for the general public than topics about, for example, public works. To simplify the usage and to raise the interest in governmental information, there exists the proposal to unify all services into one web portal (Al-Yafi et al., 2014) like the government started with Hukoomi—"the official gateway to all the information and services you need to live, work or stay in Qatar" (Hukoomi, 2016, 2nd paragraph). Until now, the maturity level of this e-Government service is rather low, but the plans and promises of the country's e-Government Strategy 2020 give rise to the hope of a successful implementation in the future. The purpose will probably increase the social media popularity of this overall e-service portal, but maybe to the detriment of the now successful institutions. A huge change of participation can certainly only be expected, when Qatar's inhabitants support their government by posting content, asking questions and requesting information themselves. Indeed, the inhabitants participate in discussions when the governmental institution starts the dialogue.

The support by the citizens to establish a fruitful government-citizen relationship is therefore only partly present nowadays. In general, the e-Government sector needs more promotion so that the citizens will truly benefit from it, even if the government is already on a right track today. What is completely missing is the incentive for the citizens to start a conversation by themselves. The mutual support via social media channels would result in the involvement of citizens in the development plans of the country. Are there any other specific groups, like scientists, engineers or political leaders, or cultural factors playing a significant role in the implementation? Does the building-up of a knowledge-based economy work with their support? Will the future workforce be well prepared for the challenges it will have to meet?

References

Al-Khouri, A. M. (2013). E-government in Arab countries: A 6-staged roadmap to develop the public sector. *Journal of Management and Strategy, 4*(1), 80–107.

Al-Shafi, S., & Weerakkody, V. (2009). Understanding citizens' behavioural intention in the adoption of e-government services in the state of Qatar. *Proceedings of the 17th European Conference on Information Systems* (pp. 1618–1629). Verona, IT: EICS.

Al-Shafi, S., & Weerakkody, V. (2010). Factors affecting e-government adoption in the state of Qatar. *Proceedings of the European, Mediterranean and Middle Eastern Conference on Information Systems: Global Information Systems Challanges in Management* (pp. 1–23). Abu-Dhabi, AE: EMCIS.

Al-Yafi, K., Osman, I. H., & Hindi, N. M. (2014). Exploring user satisfaction of the public e-services in the state of Qatar: Case of traffic violations e-service provided by the Ministry of Interior. *Twentieth Americas Conference on Information Systems, Savannah.*

Bonsón, E., & Ratkai, M. (2013). A set of metrics to assess stakeholder engagement and social legitimacy on a corporate Facebook page. *Online Information Review, 37*(5), 787–803.

EL-Haddadeh, R., Weerakkody, V., AL-Shafi, S. H., & Ali, M. (2010). E-government implementation challenges: A case study. *Proceedings of the Sixteenth Americas Conference on Information Systems* (pp. 1–14). Lima, PE: AIS Electronic Library (AISeL).

Fietkiewicz, K. J., Mainka, A., & Stock, W. G. (2017). eGovernment in cities of the knowledge society. An empirical investigation of Smart Cities' governmental websites. *Government Information Quarterly, 34*(1), 75–83.

Hukoomi. (2016). *Welcome to Hukoomi.* Retrieved from www.gov.qa.

ictQatar, (2014). *Qatar E-Government 2020 Strategy*. Doha, QA: Ministry of Information and Communication Technology.

Kostopoulos, G. K. (2004). E-government in the Arabian Gulf: A vision towards reality. *Electronic Government, 1*(3), 293–299.

Mainka, A., Hartmann, S., Stock, W. G., & Peters, I. (2014). Government and social media: A case study of 31 informational world cities. *Proceedings of the 47th Hawaii International Conference on System Sciences* (pp. 1715–1724). Washington, DC: IEEE Computer Society.

Mainka, A., Hartmann, S., Stock, W. G., & Peters, I. (2015). Looking for friends and followers: A global investigation of governmental social media use. *Tranforming Government: People, Process and Policy, 9*(2), 237–254.

Mansour, A. M. (2012). E-government in the gulf cooperation council countries: A comparative study. *Journal of the Social Sciences, 40*(1), 11–43.

Weerakkody, V., El-Haddadeh, R., & Al-Shafi, S. (2011). Exploring the complexities of e-government implementation and diffusion in a developing country: Some lessons from the state of Qatar. *Journal of Enterprise Information Management, 24*(2), 172–196.

Chapter 8
Knowledge-Based Development

Abstract In Qatar, political decision making is mainly up to the rulers without them receiving advice from knowledge-workers. Decisions are made with regard to religion and other circumstances like the development of the country's wealth. Especially the development of knowledge-intensive companies and of the K-12 education system—which breeds the future leaders and decision makers of the country—has to be planned and funded by the leaders.

Keywords Qatar · Knowledge-based development · Moslem religion
Knowledge society · Knowledge-based companies · Knowledge-intensive start-up
Education · K-12 · Schools

With the change from an oil-based to a diversified economy, Qatar tries to enhance its global identity as it aims at becoming a regional hub for education, knowledge and culture (Alraouf, 2011). To realize this goal, large steps were taken by investing in knowledge-based urban development while implementing major projects like Education City, Qatar Science and Technology Park and the Museum of Islamic Art (Alraouf, 2012). Qatar's political leaders are aware of the fact that, besides providing appropriate facilities, the country is in need of innovation to advance economic diversification. Thus, one further ambition is to establish a leading environment for innovation and entrepreneurship (QF, 2013). This overall modernization of Qatar seems to be a challenge given the rather conservative society of the country. However, as declared in the National Vision of the country, all improvements should be made by preserving the local culture and traditions (GSDP, 2008).

8.1 Support and Restrictions

While discussing master plans and other decisions concerning changes in the society, the political and administrative elites of Qatar do not systematically consult scientists or engineers. "The reins of government are tightly held in the hands of the

J. Gremm et al., *Transitioning Towards a Knowledge Society*,
https://doi.org/10.1007/978-3-319-71195-9_8

ruling family which usually makes decisions on every major issue with minimal popular input into the decision-making process" (Nakhleh, 2014, p. 162). Sometimes scientists are invited by politicians or are visited by them in the universities, but this does not occur on a regular basis. Furthermore, this communication happens mostly between people with higher position levels, like the director of a university or a research institution. As Qatar has not developed this kind of culture, yet, the political elites are not really interested in the advice of the faculty of universities or other higher education institutions (Interview partners 1, 12, & 13). Since "political decision making in Qatar is a highly centralized process that involves no more than a handful of individuals" (Kamrava, 2015, p. xi), every ministry and every decision maker has its or his own well-educated consultant who is entitled to give advice. They are, however, not connected to the universities (Interview partners 7 & 9).

As the first imperative of the Moslem religion is "read!", our interview partners are strongly of the opinion that the Islam does not stand in the way of knowledge-based development, but that it actually supports the establishment of a knowledge society (Interview partners 5, 9, & 13). "Even though Qatar is [...] a Muslim country, it has no religious institutional attachments that dictate or influence its internal affairs" (Yamani, 2006, p. 5). The country is therefore not constrained when it comes to implementing decisions regarding knowledge-based development. However, as interviewee 1 mentions, whether religion does support this kind of development or not, is strongly dependent on the interpretation of the Moslem religion by the individual. "Although innovation in science and technology is encouraged in Islam, for many the word carries a connotation of overthrowing or contradicting established tradition" (Weber, 2011, p. 2595). While religion does not oppose knowledge-based development and innovation, the rentier mentality of Qatari nationals does. When it comes to the Arabian culture, there is no evidence that it does not support knowledge-based development. Centuries ago, the Arabian culture was quite open towards knowledge. The tribes in Qatar are not at all against education, but rather against the large number of foreigners in the country. In fact, reading and thinking are important aspects of the culture even though that option is not generally accepted in practice (Interview partners 1, 4, & 7).

All of our interviewees agree that the political leaders adequately fund the emerging knowledge society. Large investments have been made to finance education, science, technology, innovation and telecommunication infrastructure as well as services (Tadros, 2015; Fig. 8.1).

The large majority of foreign talents is brought to the city primarily for the reason to support knowledge-based development. It is due to the wealth of the country that Qatar has the opportunity to establish a knowledge society. However, this funding is strongly linked to the oil price. Lately the price has gone down and as a consequence the amount of money that is invested into the establishment of the knowledge-society is reduced. Qatar tries to be more aware of the money it spends and has cut lots of budgets that were once dedicated to fund the knowledge society. This is reflected in the reduction of the funding of knowledge-based institutions. For instance, the funding of universities has been reduced by 10%. Expatriate scientists

Fig. 8.1 Large investments in education and the corresponding infrastructure

have been dismissed resulting in them being compelled to leave the country within one month. Moreover, due to this price reduction, it has become more difficult to receive funds for research projects. The total amount of available research fund has gone down and there are more restrictions now, such as only specific areas of research are privileged to receive any funds (Interview partners 4, 7, 9, & 12). Already since 1995, with the inauguration of a new Emir, the country tries to breed more well-educated people. However, even though a lot of money is spent on the change of society, the results hereof are still sparse. The number of students and the scientific output of Qatar need to be further improved (Interview partner 7).

8.2 Development of the Corporate Landscape in the Knowledge Sector

When regarding the opinions of our interviewees, it becomes obvious that they do not think of Qatar as a country where there will exist a lot of knowledge-based companies with workplaces for knowledge workers in the near future. This is in line with the findings of the Global Information Technology Report 2016, which reveals that workers in the knowledge sector are a rarity in Qatar. In 2016, only 18% of the country's total workforce are employed in knowledge-intensive jobs which include all managers, professionals and technicians (Baller, Dutta, & Lanvin, 2016). Qatar is aware of the need to increase the contribution of its GDP to knowledge-intensive industries, which include ICT, R&D, specialty chemicals, aerospace, medical devices and pharmaceuticals, but not, for instance, oil and gas or financial services. The fixed target is to increase the GDP contribution from 4% to more than 10% in the next ten to twenty years. This would not only result in the increase of the amount of knowledge-intensive companies in the country, but also of that of the number of workplaces for knowledge workers (QF, 2013). Currently there is still a lack of investment in these companies (Interview partner 12). Especially when it comes to high-tech industries and consultants, interviewee 1 is dubious about their sufficient availability in the future. With the introduction of Qatar's development strategies, the number of knowledge-based firms, like advanced producer services and high-tech companies, has increased, but many offices were opened only for a certain amount of time and the headquarters of these knowledge-based enterprises remain outside of the country (Conventz, Wiedmann, & Salama, 2015). For financial services, this seems to be quite the contrary though. Primarily in Islamic banking, Qatar's financial services are of good quality. Many of Qatar's banks that are based on the Islamic banking system are very successful (Interview partners 1 & 12). The Islamic banking system bases on risk-and-reward sharing instead of operating with interest and deposits (Iqbal & Mirakhor, 2011). Since this system is in the act of becoming popular around the world, our interviewees are confident that there will be enough outstanding financial services in Qatar in future (Interview partner 13). In Qatar's Science and Technology Park, all companies are foreign. Since the main aim of the companies located there is to generate profit, they do not primarily employ Qataris, but rather workers with the most knowledge and skills. Therefore, Qataris do not get into the higher positions of knowledge-based companies located in the country. For foreign employees it is difficult to establish new knowledge-intensive companies due to the regulatory framework of Qatar. The sponsorship system is hindering talented foreigners to build up high-tech institutions. As the working contracts are restricted, there is no chance for expatriate workers to change anything in the country (Interview partners 4, 5, & 8).

According to our interviewees, there are many start-ups in Qatar (Interview partners 1 & 13). Being aware of the importance of innovation, which is highly present in clusters of smaller entrepreneurial organizations, Qatar focuses on supporting start-ups in niche areas (QF, 2013). Furthermore, the country tries to encourage young residents to start up their own companies by offering courses for entrepreneurship at Qatar University (Interview partner 9). There is some further support of start-ups beside that: For instance, the Qatar Business Incubation Center (QBIC) tries to encourage and support entrepreneurs to establish and develop their own businesses. They work with start-ups as well as with scale-ups and pursue the objective to increase the number of profit-yielding companies in the country. Thereby they keep to the goal of the National Vision to strengthen and to diversify Qatar's economy (QBIC, 2016). When it comes to knowledge-intensive start-ups, the same problems arise as with knowledge-based companies: Every idea needs to be sponsored by a Qatari national unless the idea comes from a Qatari himself.

However, it is still rare that a Qatari national implements his own ideas by founding an enterprise (Interview partner 1). In general, there is a low level of entrepreneurship in Qatar. The number of companies with less than 100 employees in Qatar's non-oil sector is very small, with a growth rate of only 3% per year (QF, 2013). This might indicate that the number of start-ups in the country is increasing, albeit not significantly.

8.3 Development of K-12 Education

It is a challenge for the country to have enough well-trained teachers in elementary and high schools in the future. Due to the large amount of expats that come to the country, Qatar's population is growing and with it the number of children in need of education is increasing (Interview partner 5). Qatar has long since recognized the need to improve its weak public school system. In 2001, the country started to implement an educational reform. However, even though the plan to transform ministry schools to independent schools has been accomplished within a short timeframe, the reform did not bring the anticipated success (Alkhater, 2016). Teachers of independent schools did not get enough time to learn the new style of teaching and to familiarize with the new curriculum and teaching materials. As a result, they are ill-prepared and incapable to teach their students properly. Especially Qatari students often lack the skills to enter postsecondary education (Zakhidov, 2015). Today, most of the teachers in elementary and high schools are foreigners. This is why Qatar needs to establish adequate working conditions for expatriate teachers to come to or to stay in the country (Interview partner 10). Interviewee 1 is unsure if well-trained teachers continue to seek work in Qatar because the country has decided that all ministries should communicate in Arabic. This might make expatriates feel unwelcomed. Another problem might be the implementation of Qatarization in the country. Well-educated expats are dismissed by companies to be replaced by Qataris who might be less qualified. This might be one of the counter-arguments of qualified foreign teachers to come to Qatar (Interview partner 8). There are differences between teachers in public and those in private schools. While teachers in private schools get a lot of training, teachers in public schools do not receive the same chances to develop their potential (Interview partner 13). If the money is available, parents prefer to send their children to private schools (Interview partner 8). "The quality of education at independent schools, parents say, lags behind international and private schools" (Paschyn, 2013, 5th paragraph). Especially expat students are enrolled in private schools. Qataris are less frequently sent to international schools in Qatar. They do graduate from high school; however, they do not receive the best grades. Altogether, only 10–15% of all students are enrolled at a university after high school graduation each year (Interview partners 10 & 12).

In Qatar, there is a push towards education and this passes on to parents who want to see their children well educated. Expatriates are still more eager than

Qataris to enroll their children in the best education institutions available. If they have enough money, parents send their children to private schools. Especially expat students are enrolled in those institutions. The number of Qatari students enrolled in international schools in Qatar is much lower (Interview partners 1, 8, 12, & 13). However, the "popularity of private schools is rising among Qataris, who are increasingly paying from their own resources to send their children to international schools" (GSDP, 2011, p. 14). When it comes to Qatari females, the willingness to be educated often comes from the students themselves, rather than from their parents (Interview partner 1). As for the males, they are not as eager to study as the females and are often sent to the military if their parents are not behind education. The majority of parents and grandparents in Qatar has not been to university, so that their children are the first generation to receive higher education. The motivation of parents and students depends on the level of education: At the lower levels, parents send their children to school just to have them do something. Thus, there are often attendance problems in elementary and high schools. Parents who send their children to university or other higher education institutions want them to become engineers or directors. The only ideal for the students encouraged by their parents to attend universities is to go to the top (Interview partner 12). Mostly the attitude of the parents is adopted by their children.

> One of the deep-rooted strengths of the GCC culture is being able to be inspired and following role models. GCC youth look up to and take direction in particular from two groups: their family and community leaders. Whether a family father encourages his children to go for a government job because it is guaranteed and safe or whether he encourages his children to study hard to get a good private sector job because it provides greater opportunities in the long run, will have a big impact on the decisions his children will take (Hoteit, Tuerpitz, & Luelsdorf, 2015, p. 4).

8.4 Summarizing Qatar's Support of Knowledge-Based Development

The political leaders of Qatar are willing to develop the oil dominated society into a knowledge-based and diversified society. They are financing this transition adequately by investing large amounts of money into knowledge-based urban development in the fields of education, science, technology and research. While Qatar is a rather conservative society that keeps closely to the traditions of the Arabian culture and the rules of the Muslim religion, there is no evidence that culture and religion principally stand in the way of knowledge-based development. However, due to the tradition that the ruling family does not include other people when making major decisions, knowledge workers like scientists and engineers are not consulted as a source for well-founded knowledge when discussing master plans and other important societal changes, even though the country might benefit from their suggestions. Some of the investments that were made and measures that were taken at the instruction of the political leaders have not yet brought about the anticipated success. Even though Qatar tries to encourage entrepreneurs to start up new or scale up their existing companies in the country, the amount of enterprises and start-ups in the knowledge-sector and with that the number of jobs for knowledge workers is still strikingly low and also not expected to grow significantly in the near future. The measures that were taken to improve Qatar's education system in order to give young Qataris the chance to acquire the needed knowledge early in life so that they become later part of the talented workforce of Qatar have also not succeeded, yet. As shown by the 2012 PISA, the level of the students in Qatar is much lower than the average level of students in the world (Kelly, Xie, Winquist Nord, Jenkins, Ying Chan, & Kastberg, 2013). However, even though the state is constantly ranking among the least academically proficient countries, the results of Qatar are tending upwards (Cheema, 2014). The students' performance depends strongly on the expertise of the teachers. Some of the teachers in Qatar are insufficiently qualified to adequately educate the students in the country. Especially the teachers in public schools are lacking skills and training opportunities. Most of the teachers in the country are non-Qataris which does not seem to change in the near future. If the country wants to assure that there are enough well-trained teachers in the future, it needs to create incentives for excellent foreign teachers to come while providing appropriate training to all the existing teachers. However, the students' performance does not only depend on the skills of the faculty, but also on their own motivation to be successful in school, what again depends on the encouragement of their parents. With the new awareness of the importance of knowledge in the country, parents are more motivated than before to see their children well educated. However, since most of the parents and grandparents of the student body have not been to university themselves, some of them do not see the importance of encouraging their children to work hard and become part of the elite workforce of the country.

Has Qatar, after it has introduced a reform to change its K-12 education system, taken further remarkable measures to improve its higher education landscape? Did the political elites make large investments in the establishment of sensational higher education institutions to enhance the quality of Qatar's higher education system or did they not recognize the need to introduce a large-scale transformation? Which higher education institutions are present in the country? What are the characteristics of the faculty, the students and the graduates?

References

Alkhater, L. R. M. (2016). Qatar's borrowed K-12 education reform in context. In M. E. Tok, L. R. M. Alkhater, & L. A. Pal (Eds.), *Policy-making in a transformative state. The case of Qatar* (pp. 97–130). London, UK: Palgrave Macmillan.

Alraouf, A. A. (2011). Assembling identity for an emerging urbanity: Doha, Qatar as a creative city. In *Proceedings of The City: 2nd International Conference—Culture, Society, and Technology.* (pp. 48–63) Vancouver, British Columbia, Canada, May 4–6, 2011.

Alraouf, A. A. (2012). Dohaization: constructing a new urbanity brand amid Gulf and Middle Eastern cities. Retrieved from www.analchemyofarchitecture.blogspot.com.

Baller, S., Dutta, S., & Lanvin, B. (2016). *The Global information technology report 2016: innovating in the digital economy*. Geneva, CH: World Economic Forum.

Cheema, J. R. (2014). The migrant effect: an evaluation of native academic performance in Qatar. *Research in Education, 91,* 65–77.

Conventz, S., Thierstein, A., Wiedmann, F., & Salama, A. M. (2015). When the oryx takes off: Doha a new rising knowledge hub in the Gulf-region? *International Journal of Knowledge-Based Development, 6*(1), 65–82.

GSDP. (2008). *Qatar national vision 2030*. Doha, QA: General Secretariat for Development Planning.

GSDP. (2011). *Qatar national development strategy 2011–2016*. Doha, QA: General Secreteriat for Development and Planning.

Hoteit, L., Tuerpitz, A., & Luelsdorf, L. (2015). *Motivation and perseverance: the missing link to gcc knowledge economies*. Dubai, AE: The Boston Consulting Group.

Iqbal, Z., & Mirakhor, A. (2011). *An introduction to islamic finance: theory and practice* (2nd ed.). Singapore, SG: Wiley & Sons.

Kamrava, M. (2015). *Qatar: small state, big politics*. Ithaca, NY: Cornell University Press.

Kelly, D., Xie, H., Winquist Nord, C., Jenkins, F., Ying Chan, J., & Kastberg, D. (2013). *Performance of U.S. 15-year-old students in mathematics, science, and reading literacy in an international context: first look at PISA 2012*. Washington DC: National Center for Education Statistics.

Nakhleh, E. A. (2014). Political participation and the constitutional experiments in the Arab Gulf: Bahrain and Qatar. In T. Niblock (Ed.), *Social and economic development in the Arab Gulf* (pp. 161–196). Oxon, UK: Routledge.

Paschyn, C. M. (2013, October 25). Zig-zagging education policies leave Qatari students behind. *Al-Fanar Media*. Retrieved from www.al-fanarmedia.org.

QBIC. (2016). *This could be you: Be the next Qatar success story*. Doha, QA: Qatar Business Incubation Center.

QF, (2013). *Qatar foundation strategic plan 2013–2023*. Doha, QA: Qatar Foundation.

Tadros, M. E. (2015). *The Arab Gulf states and the knowledge economy: challenges and opportunities*. Washington, DC: Arab Gulf States Institute.

Weber, A. S. (2011). The role of education in knowledge economies in developing countries. *Procedia-Social and Behavioral Sciences, 15,* 2589–2594.

Yamani, S. (2006). Towards a national education development paradigm in the Arab world: a comparative study of Saudi Arabia and Qatar. *Al Nakhlah*. Retrieved May 31, from www.fletcher.tufts.edu.

Zakhidov, D. (2015). K-12 education in Qatar. In K. C. Ulrichsen & A. Marnicio (Eds.), *The 2015 Doha conference: findings from the student-led public diplomacy program* (pp. 9–13). Doha, QA: Rice University's Baker Institute for Public Policy.

Chapter 9
Universities and Other Institutions of Higher Education

Abstract The higher education sector of Qatar depends heavily on foreign knowledge. This knowledge is imported by foreign teachers or even by branches of international high-class universities that are unified in a prestigious campus called Education City. These facts set the stage for research and education on a Western level so that the young Qataris' only need to grab their chance. Apart from the exceedingly good higher education facilities brought by the foreigners into the country, they also bring their culture and lifestyle with them, what is in turn not well received by many Qataris.

Keywords Qatar · Higher education · Universities · Education City
Qatar Foundation · Qatar University · Research funding · University funding
Gross domestic expenditure on research and development · Internationality
Branch campuses

Qatar already recognized the need of successful knowledge societies to build up a fluctuating knowledge-based economy. In times of the continuously rising importance of knowledge accompanied by the development of new knowledge, researchers and scientists play a significant role in forming a pool of talent of which organizations and industries can make use to remain competitive (QF, 2013). "Higher education is viewed an essential to become a knowledge-based society and plays a central role in establishing and developing Qatar as an education hub" (Ibnouf, Dou, & Knight, 2014, p. 44). Based on this idea, the former Emir and his consort Sheikha Moza "have initiated the development of state-of-the art educational and research facilities in order to create a sustainable foundation for a more diversified economy" (Salama & Wiedmann, 2013, p. 95). The Qatar Foundation, established for this purpose, relies on three pillars: education, science and community development, where education is the "key instrument for modernizing the country, reducing the dependence on oil and gas and reinforcing the international competitiveness of Qatari citizens" (Ibnouf et al., 2014, p. 44). This idea bases on the Qatar National Vision 2030 that describes the QF's and with that the country's plan of having a lasting effect on the human, social and economic development

J. Gremm et al., *Transitioning Towards a Knowledge Society*,
https://doi.org/10.1007/978-3-319-71195-9_9

through education and research (GSDP, 2008). Nowadays QF houses some of the world's greatest universities that bring altogether over 3000 researchers and faculty members along. Unified in the Education City campus, they should attract students not just from the surrounding countries, but rather from the whole world and consequently bring new talents to Qatar (QF, 2013; Salama & Wiedmann, 2013). Under the umbrella of the country's own Hamad bin Khalifa University (HBKU) this association has "set a new standard for higher education in Qatar and the region" (QF, 2013, p. 2) by now. Facing the challenge of becoming a knowledge hub, the country has all the attributes to develop expertise for the future post-oil economies (Baghdady, 2016). Qatar's higher education and research policy pursues two approaches, namely "(1) to match the strongest global exemplars through borrowing and direct importation of existing organizational capacity, faculty and staff, and accumulated reputation; and (2) to cultivate native human capital through massive infrastructure investments and development of educational structures" (Powell, 2014, p. 255).

9.1 The University Landscape of Doha

As a factor of fundamental importance that paves the way for Qatar to become a knowledge society, Education City comes to the fore. Until its establishment, which started in 1998, "Qatar University was the only higher education institute in the country" (Ibnouf et al., 2014, p. 44) and "has been the focal point of discussion and decision-making" (Abu-Tineh, 2011, p. 636) in terms of knowledge. Education City's non-profit status—which means that there is no third party interested in the financial affairs and the sponsorship is provided exclusively by the government—isolates the organization from most of the other branch campuses in the region. Some critical voices note that the very location outside the boundaries of the city let arise some kind of disconnection between the international branch campuses and the city or the country—the universities are "not coupled with the normal society" (Brandenburg, 2012, p. 239). On the contrary, Qatar University is more integrated, but regarding the holistic attention too often neglected (Mitchell, 2014). In spite of all concerns, Qatar is a firm believer in Education City that includes more than just academic programs namely research initiatives as well as science and research institutions, like Qatar Science and Technology Park and Sidra Medical and Research Center, in notable architectural buildings (Ibnouf et al., 2014; Salama & Wiedmann, 2013; Figs. 9.1 and 9.2).

And as soon as the railway project, which will among others "run from the industrial areas in the south through the center of Doha and to the Education City" (Shaaban & Radwan, 2014, p. 246), is finished, it will ease the arrival for everyone and the location of Education City will no longer be a problem. During our stay in spring 2016, the project was, however, still under construction.

Fig. 9.1 Prestigious architecture in Education City, here: Georgetown University Building in Qatar

Fig. 9.2 Prestigious architecture in Education City, here: Texas A&M in Qatar

Education City set itself the goal "[t]o effectively prepare academically superior students to assume responsible and high-ranking professional and leadership positions in Qatar and beyond" (CMU, 2016a, 8th paragraph). To fulfill this purpose was an incentive for different universities to come to Qatar so that today Education City houses overall eight foreign universities or colleges and one state-owned university—all owned by Qatar Foundation or the country (QF, 2015a; Interview partner 4):

- Virginia Commonwealth University in Qatar
- Weill Cornell Medical College in Qatar
- University College London Qatar
- Texas A&M University at Qatar
- Carnegie Mellon University in Qatar
- Georgetown University—School of Foreign Service in Qatar
- Northwestern University in Qatar
- HEC Paris in Qatar
- Hamad Bin Khalifa University (state-owned).

With this assortment, the country built up a high-class campus that houses without exception private universities that are under "the top 10 or 15 institutions in the world in the particular field of study" (Ibnouf et al., 2014, p. 50), "each to specialize in a different discipline" (Wilkins, 2011, p. 75). The selection of programs on offer is chosen "to prepare Qatari and expat talents for the workforce and future needs of the country" (QF, 2013, p. 22). The Virginia Commonwealth University that offers degrees in fine arts educates, among others, for the news agency Al Jazeera and the cultural village Katara. The academic healthcare training by Weill Cornell Medical College is launched to train for the new medical center Sidra and for the hospitals of Hamad. Texas A&M University concentrates with its different engineering programs on graduates for the traditional oil and gas sector. In addition, the financial sector is covered by graduates of Hamad Bin Khalifa University (Table 9.1).

Moreover, if a company develops a new field, after a while new courses are firstly generated and then offered by the universities (Interview partner 6). By establishing these universities and courses in the country, Qatar is now able to serve the market with self-trained workers. Qatar University (established 1973) (Fig. 9.3) indeed covers most disciplines (Ibnouf et al., 2014)—since fall 2015 also medicine (College of Medicine, 2016)—but now foreign expertise helps improving what the public university cannot accomplish alone.

The continuous comparison between Education City and Qatar University leaves its marks and spreads the feeling that too much attention is paid to the branch so that the national university fades into the background (Mitchell, 2014). However, it was not the intention: With the import of knowledge, foreign education has been implemented in the country to focus increasingly on the global economy, to develop the competitive ability and to be internationally recognized (Vora, 2014). "[T]hinking critically and articulating opinions, honing oral and written communication skills, emphasizing lifelong learning, and using a variety of tools and resources" (Ghabra & Arnold, 2007, pp. 2–3) are the basic pillars on which the new education relies. Sheikha Moza still sees further potential in this idea and tries to promote the cross-cultural dialogue and the international understanding by being multicultural, multireligious and multiethnic (Ibnouf et al., 2014).

Most of the institutions in Education City offer mainly Bachelor degrees, less Master degrees and still less Ph.D.'s (Interview partner 12). At Qatar University, you can achieve all three graduate statuses (QU, 2016a, c).

Table 9.1 Universities in Education City and their programs, the industry they train for and typical companies the graduates will work for

University	Study program	Industry	Typical company
Virginia Commonwealth University in Qatar (1998)	B.A. Fashion design, B.A. Graphic design, B.A. Interior design, B.A. Painting and printmaking, B.A. Art history; M.A. Design	Art and design	Al-Jameela Magazine Al Jazeera Arteligno Interior Designers Katara Cultural Village many start-ups
Weill Cornell Medical College in Qatar (2001)	Pre-Medical/Medical Education (M.D.)	Healthcare	Hamad Medical Corp. Sidra
University College London Qatar (2012)	M.A. Library and information studies, M.A. Museum and gallery practice	Librarianship Museums	Qatar National Library Qatar Museums
Texas A&M University at Qatar (2003)	B.S. Chemical engineering, B.S. Electrical and computer engineering, B.S. Mechanical engineering, B.S. Petroleum engineering, M.S. / M. Eng. Chemical engineering	Chemical engineering Electrical and computer engineering Mechanical engineering Petroleum engineering	Qatar Petroleum Maersk Oil Qatar Qatar Shell RasGas Qatar Gas
Carnegie Mellon University in Qatar (2004)	Biological sciences, Business administration, Computational biology, Computer science, Information systems (all Bachelor's degree)	Biology Business administration Information and communication	Vodafone Microsoft Qatar Petroleum Ooredoo ictQatar Qatar Airways
Georgetown University—School of Foreign Service in Qatar (2005)	B.S. Foreign service (Culture and politics, International history, International politics, International economics), M.S. Foreign service, M.A. Arab studies, M.A. German and European studies, M.A. Latin American studies,	Foreign service Politics	Government Al Jazeera Qatar State Security Qatar National Bank

(continued)

Table 9.1 (continued)

University	Study program	Industry	Typical company
	M.A. Global, international and comparative history		
Northwestern University in Qatar (2008)	B.S. Communication, B.S. Journalism	Communication, media Journalism	Al Jazeera Qatar Shell
HEC Paris in Qatar (2010)	Executive MBA, MBA Strategic business unit management	Management	All companies
Hamad Bin Khalifa University (2012)	B.S. Computer engineering, M.A. Islamic studies, M.S. Islamic finance, M.A. Translation studies, M.A. Audiovisual translation, M.S. Data science and engineering, M.S. Cybersecurity, M.S. Public health, M.S. Energy and resources, doctorate law, M.S. / Ph.D. Biological and biomedical sciences, Ph.D. Computer science and engineering,	Islamic Studies Islamic Finance Translation Computer Science Energy Law Healthcare	Qatar National Bank Ooredo Qatar Petroleum Hamad Medical Corp. Sidra Qatar Biomedical Research Institute Qatar Islamic Bank

(continued)

Fig. 9.3 Qatar University: gorgeous architecture and landscape design

9.2 Level of Research and Education

Compared to the Western examples, the research quality in Education City is just as good, but in quantity the universities cannot match up to them (Interview partner 1). The pressure on the institutions and their faculty is high so that QF pushes the standard indirectly to a quite good level. At Qatar University, the standard seems to be just as high. Taken as a whole, the research output increased steadily during the last years. Due to the higher number of researchers, Qatar University naturally produces more than the universities in Education City. What they publish is despite all that quite good (Interview partners 4, 6, & 13). Nevertheless, in every case the quality and quantity of research depends on the individuals—a department is either productive or unproductive (Interview partner 6)—like surely in every university.

Concerning the level of education, it looks similar: Courses and degrees are the same as at the foreign universities' home campuses and every paper, thesis or dissertation is revised by a lecturer from there. Moreover, the teachers' quality follows the international standard. Exclusively people that are qualified to work at the home campus are employed at the Qatari branch. In relation to Western standards the time that has to be calculated for teaching is considerably higher even if some students—that are still the best—have to complete an Academic Bridge Program successfully before the enrollment. These courses are supposed to improve the rather poor qualifications of the students. However, in the end of their studies, students achieve a comparably good level (QF, 2013; Interview partners 1, 4, & 10). Due to the rather small number of students in Education City's universities, every student gets more attention—teachers have about six hours per day to spend with a student. At Qatar University as well as at most of the universities in other countries, there are simply too much students for such an individual support so that the standard of education compared to Education City and the Western universities is described as rather low. This is further complicated by the fact that the teaching language is Arabic, a language that many people are not able to speak, which in turn reduces the number of potential lecturers. In addition, the missing range of documents in Arabic is problematic (Interview partners 1 & 13). It should be added that teachers of the branch universities do not stay there for a long time—"they are just on vacation" (Interview partner 8)—which is apparently no problem because they of course do not leave until the end of a semester. Emerging cities are overall characterized by short-term stays, and usually do not become the home of the expats, who are in most cases there to just earn money and to gain experience (Salama & Wiedmann, 2013).

9.3 Research and University Funding

Support for the branch campus project (and certainly other educational projects at Qatar University) is provided by the government—at times they are more generous, at some less—depending on the oil and gas prices. Currently, money is saved everywhere, but nevertheless the universities benefit from the investments and are still satisfied (Interview partners 1, 6, & 8).

The funding of research in Education City is deemed to be attractive (Ibnouf et al., 2014), which is also confirmed by our interview partners. The whole lot of them is satisfied, there is nothing to beef, the budget for training and research is generous, travels and conferences are paid (Interview partners 1, 4, 8, & 12). In terms of resources, Qatar University gets, according to our interview partner 1, nevertheless more resources and research is in total better funded. The branch universities in Education City sometimes engage specialized lecturers that come from the home campus, other universities or companies which likewise speaks for sufficient possibilities of use of resources (Interview partner 12). The funding is overall very satisfactory. The only circumstance that is judged to be not as satisfactory as it could be is the organizational approach. In this context it is important to say that faculty members are positioned as broadly as possible—one specialist for each department (Interview partner 4). However, is there anything else that attracts premium teachers from all over the world to come to Qatar? In general, people coming from abroad do not have to pay income taxes, neither to Qatar nor to their home country. The situation is different for US-Americans and Australians: They have to pay to their home state, but not to Qatar. In fact, these people are appointed in the USA or in Australia—in the country were the branch campus is originally home—and have to deliver their statutory levies. Due to this appointment, another advantage stands out: If the researchers want to change their task and maybe also their placement, they can easily go (assuming a job vacancy) to the university's home campus and work there (Interview partners 1 & 8). The other way around, interview partner 1 resigned from his ten-year position just to come to Qatar for three years because of attractive salary and pleasant conditions (Fig. 9.4). To

Fig. 9.4 Many expatriate talents live at The Pearl where several restaurants and cafes make life comfortable

become rich is despite all that impossible. If expats even bring their family to the country, they have to bear the high costs for the country's best schools that shrink the salary significantly (Interview partner 8).

From the students' view, the development potential at Qatar's universities—regardless of whether public or private—is great, because they have vast opportunities to travel around the world which makes some of the teachers become jealous (Interview partner 12).

Indeed, both educational institutions—Education City and Qatar University—are financed by QNRF (QU, 2016b; Interview partner 10). QNRF provides two funding programs that both educational institutions benefit from: (1) an Undergraduate Research Experience Programme (UREP) that "will fund student research projects at universities and colleges in Qatar as well as providing a mentoring programme" and (2) a National Priorities Research Programme (NPRP) that "is aimed at research in academia, industry and the public sector" (Küster, 2010, p. 17). The NPRP is thereby with grants between 20,000 U.S. dollar and 350,000 U.S. dollar per year the biggest education funding program in the country.

The Gross Domestic Expenditure on Research and Development—a common indicator for expenditures on R&D measured per Gross Domestic Product of the country—ranges, despite the varying prices, since a few years consistently at a value of 2.8 (2013) or 2.7 (2014) percent (Industrial Research Institute, 2016). The forecasted values for the next years remain constant as well. The entirely adequately funded country achieved almost the as "good" referred value of 3.0%. Among the top 40 R&D spending countries we find only Saudi Arabia as a second country from this region. Its GERD per GDP value (0.32%) is even so just a fraction of Qatar's expenditure. Qatar's 2014 values are comparable to those of the United States (2.78%), Germany (2.85%) and Denmark (2.90%)—what is remarkable due to the small number of inhabitants. Highest values are measured for South Korea (3.60%), Finland (3.50%), Sweden, and Japan (both 3.40%) (all values from the year 2014; Industrial Research Institute, 2016). "Most Middle East countries will experience strong GDP growth in 2014, but are constrained by weak R&D infrastructure—with exceptions such as Israel and Qatar" (Battelle, 2014, p. 4) whose R&D investments are competitive to global prevailing levels. Overall 41.9% of the Qatari expenditure is spent on the higher education sector—the others for business and government (The Peninsula, 2015). The budget is then again not just spent on the bricks and mortar of Qatar's education like buildings or computers—the country additionally uses QNRF as a funding body that supports research projects that are best fit for the National Vision (Mitchell, 2014). It should be added that beyond the relatively high expenditure on R&D no university in Qatar has to pay any taxes meaning that the budget does not shrink as a result of these costs (Interview partner 4).

9.4 Internationality: Opportunity or Risk?

With the highest concentration of branch campuses worldwide (Becker, 2010), the Arab Region has to face the challenge of organizing a conscientious communication, adequate teaching and living opportunities as well as a fruitful collaboration. For the universities and teachers there are both, some advantages and disadvantages in wait: Many programs are simply borrowed and just modified, QF hires the home universities chosen staff and faculty, in spite of home-branch-campus collaboration in research, the Qatari branch is most closely isolated and disconnected from the main campus and consultations between them can take time (Interview partners 4, 6, 12, & 13). It gets hard when you have to take into consideration the culture because it frustrates the teachers if they cannot do things like in London or another home campus. Community work is an essential and labor-intensive task they have to accomplish. As a consequence, the teachers know their students more personally and are able to do a more specific kind of training. Maybe it would be cheaper to bring the students to the main campus, but due to the fact that girls are not allowed to leave the country alone, this is unrealizable and therefore it is deemed necessary to bring the universities to the girls rather than the other way around (Interview partners 12 & 13). The small community inside of Education City—that acts more and more as a group—makes many new contacts for all involved people possible (Fig. 9.5). Overall the organization and exchange can in some cases be described as smooth, in others as very complicated (Interview partners 4, 6, & 13)—so that it seems to depend on the branch campus itself and the organization within its university. All these conditions, developments and manifold opportunities in Education City occur under the administration of QF (Interview partner 5).

Fig. 9.5 Evidences for the internationality in Education City

In terms of internationality, Qatar University plays also a significant role, however, in a broader meaning, namely as the most international university in the world considering staff/faculty, students and cooperation. Faculty definitely benefits from this internationality because ideas, opinions and cultures from all over the world are gathered and summed up to greater and internationally accepted concepts (Interview partner 1). Working in a multi-national team brings different perspectives to the research project, opens up access to more data and unifies different knowledge and points of view. In all probability, the research team will be conditioned by multinational influences and information will be more up-to-date. For the branch campuses and their work within the Qatari location, time management and different methods make the cooperation with the home campus complicated (Interview partners 1, 6, 8, & 9). Since research teams are rather rare at the branch universities and if they exist, they consist of only two or three people (Interview partner 4), most of the researchers do not have to face this problem. Additionally, the students like the diversity of classes for the simple and same reason that "combined local and foreign knowledge [...] produce[s] better results" (Mitchell, 2014, p. 85). The offer of courses in different languages helps improving the students' multilingualism (Interview partner 1). Cooperation, which is a very important topic for every university, is made easier by having a high language competence meaning that for nearly every language a native speaker can be found (Interview partners 9 & 11). Besides these advantages, interview partner 1 spotlights on the fact that "gender separation and being international do not fit together". Moreover, the situational condition that every new faculty member in Qatar University has to be able to speak Arabic from now on and just some colleges and researchers are still allowed to practice in English does not make a future title defense easy (Interview partners 8 & 9). Good international teachers and researchers would not come to Qatar University due to this language barrier. Just those that come from other GCC or further Arabian countries will be interested in these jobs in the future. In the Qatari context, the acquisition of foreign knowledge, people and research teams—all of whom will not stay there forever—rather shows the country's inability of creating an excellent education infrastructure from their own resources (Interview partner 10).

Due to this internationalization, the majority of Qataris is afraid of losing their identity. With English as the most common language in both, private and academic life, and the loss of the Arabic language, group formation of those speaking the same language and having the same culture is happening and a feeling of overwhelming is clearly noticeable (Davis, 2010; Sharifian, 2009; Interview partners 1, 5, & 10). The situation is thereby a little contradictory: On the one hand, Qataris are frightened of Americanization and Westernization and have become more conservative recently by paying more attention to the Islam. However, in comparison to other GCC states, like the UAE, Qatar is not that westernized. On the other hand, the country looks like an American example. Qataris appear to want the internationalization, considering that they bring Western universities to their country and that the education is fully westernized (English books, Western teaching system in international schools) which is in turn launched by the Qataris consciously

(Interview partners 4, 8, 9, & 11). They bring the knowledge into the country and with it certainly the Western culture—good things like democracy and gender equality, but also bad things like the excess of consumerism (Interview partner 1). To reduce this influence, the government tries to wind down the number of foreigners, to increase the number of native workers and to push the Islamic inner-city development by providing an Islamic architecture and city planning course of study in Education City. However, without Qatari applicants—that are simply nearly nonexistent—their project of nationalizing the workforce is hard to implement. The convenient working situation in the public sector with attractive salaries and less work does not contribute to the natives' motivation for finding a job in a university or other private sector companies. It would take long time to fill the teaching, research and staff positions exclusively with Qataris (Interview partners 4, 8, & 11). Until now, every attempt of staff Qatarization in this sector has failed and the number of expatriate faculty continuously exceeds the number of Qatari faculty—in the public, but still more in the private higher education sector (Table 9.2).

In addition, Kuwaiti and Emirati that are recruited to raise the number of Arab workers are not motivated to come to Qatar (Interview partner 10). Since the data of Table 9.2 are from the academic year 2014/2015, the numbers differ a little bit from the current: In terms of faculty, we heard of just one Qatari and one Bahraini teacher as representatives of the GCC states in Education City during our interviews (Interview partners 5 & 10). According to the statistic above, overall eight Qataris worked in a professor, lecturer or teaching assistant position during the academic year 2014/2015. Interview partner 8 told us about an amount of 6% Qataris working at Qatar University and just 3% in the College of Business and Economics—in the academic year 2014/2015 about 18% of the academics were Qataris. Despite these rather slight differences, the tendency has become clear. In contrast, in the administrations some more job positions are staffed with Qataris. Overall, the few

Table 9.2 Qatar University staff and staff of private universities and colleges by nationality and professional status (2014/2015)

Status	Qatar University (Public)			Other universities and colleges (Private)		
	Number of academics	Qatari	Non-Qatari	Number of academics	Qatari	Non-Qatari
Professor	126	22	104	147	1	146
Associate Professor	223	44	179	88	1	87
Assistant Professor	330	67	263	148	1	147
Lecturer	318	31	287	563	1	562
Teaching Assistant	213	58	155	23	4	19
Total	1210	222	988	969	8	961

Source MDPS (2015)

Qatari post-graduates rather seem to go to Qatar University, if they go to a university at all (Interview partners 1 & 5).

The tendency is clear: If Qataris work at a university, then it is almost always at Qatar University; expatriates work at both universities, but the amount is distinctly higher at the private universities in Education City. Why is the ratio of Qataris at private universities that low? If Qataris support the knowledge sector, maybe they prefer to do it in a more cultural or national way and decide upon teaching at Qatar University given that half of the employees (54.2%) are Qataris or other Arabs (QU, 2014). Interview partner 8 mentions that maybe the population is not big enough to have many Qatari Ph.D.s. However, is this really the reason for the generally small number of Qatari university employees? There rather seem to be others: If they do not study, they cannot work there; if they study, a university career may be more attractive than a career in, for instance, medium-sized companies, but in turn less attractive than not working at all, as they have too many other sources of revenue; if, however, they take a job, they do not have to study for it, but receive large amounts of money anyway (Interview partners 4, 8, & 10). It seems that in general research as a career field is commonly rather not chosen (Ibnouf et al., 2014). Especially the opportunities for women are excellent as for every newborn child they get a new nanny. Instead of taking this chance, most of the Qatari women end their careers with the delivery of a baby (Interview partners 10 & 11). Men are either way unrestricted, but do not grab the chance, probably because of convenience. A further reason could be the fact that someone from the main campus conducts the application procedure—in case of universities in Education City. For those Qataris that are educated well enough the process is maybe not that easy or confident so that their motivation is once again reduced. Actually, some Qataris do work in the administration departments of the universities. Sheikha Moza appears to be honestly willing to change the situation and the development of the country, but has not been successful until now (Interview partners 1 & 11).

9.5 Summarizing Qatar's Higher Education Ambitions

In terms of higher education, Qatar is overall well-positioned. The level of research and education as well as the quality and organization of resources are mostly comparable to Western standards at all forms of education; students particularly profit from the good teaching support in Education City and have chances to overcome their low skills; internationality and multi-national research at the branch campuses as well as at Qatar University bring more benefits than limitations to the faculty; financing of education is secured and also the financial and living conditions are good. QF accomplished the creation of a higher education infrastructure that trains graduates who in turn can fill all jobs of Qatar's economy so that both students and companies should be satisfied. Beside all these excellent opportunities, we should not forget that all these achievements could just be realized with—and based on—the import of foreign knowledge, expertise and people. And with every new arrival, a part of the Qatari culture and way of life gets lost. Not just the Americanization or Westernization of education—which was mentioned as no real problem—keeps playing on the nationals' mind. Above all, the fear of losing identity is prevalent while all mentioned benefits are from the expats' perspective. The prestigious Education City project will "have policy influences in all sectors of the society, not only in education, but also in the political, economic, social, technological, communication, and demographic structures of Qatar" (Khodr, 2011, p. 523).

It remains to be seen whether Qatar accomplishes the resulting chance with their own resources—or in other words with natives that become students and thereafter knowledge workers that undertake the expats' tasks and jobs. Otherwise, they probably never escape from the dilemma of depending on the foreign expertise and losing the cultural identity. A solid basis of universities and teachers for the development of a knowledge society has been created anyway. With it, it should not be a problem to train competent graduates. However, are there any problems for the students? Are there any differences between the requirements of Qataris and expatriates? Or of female and male students? Are the study opportunities the same regardless of studying at a public or private university?

References

Abu-Tineh, A. M. (2011). Exploring the relationship between organizational learning and career resilience among faculty members at Qatar University. *International Journal of Educational Management, 25*(6), 635–650.

Baghdady, A. (2016). Transforming Qatar's PSE: achievements and concession. In M. E. Tok, L. R. M. Alkhater, & L. A. Pal (Eds.), *Policy-making in a transformative state. The case of Qatar* (pp. 131–153). London, UK: Palgrave Macmillan.

Battelle. (2014). *2014 Global R&D funding forecast*. Cleveland, US: Battelle.

Becker, R. (2010). International branch campuses: New trends and directions. *International Higher Education, 58,* 3–5.

Brandenburg, T. (2012). *Bridging the knowledge gap: internationalization and privatization of higher education in the state of Qatar and the Sultanate of Oman*. Mainz, DE: Johannes Gutenberg-Universität.

CMU. (2016a). *Life in Education City*. Retrieved from www.qatar.cmu.edu.

College of Medicine. (2016). *College of Medicine profile*. Retrieved from www.qu.edu.qa.

Davis, C. J. (2010). *American higher education in the Arabian Gulf—a force for liberalization*. Monterey, CA: Naval Postgraduate School.

Ghabra, S., & Arnold, M. (2007). *Studying the American way: an assessment of American-style higher education in Arab Countries (Policy Focus #71)*. Washington, DC: The Washington Institute for Near East Policy.

GSDP. (2008). *Qatar national vision 2030*. Doha, QA: General Secretariat for Development Planning.

Ibnouf, A., Dou, L., & Knight, J. (2014). The evolution of Qatar as an education hub: moving to a knowledge-based economy. In J. Knight (Ed.), *International education hubs: student, talent, knowledge-innovation models* (pp. 43–61). Dordrecht, NL: Springer, Netherlands.

Industrial Research Institute. (2016). *2016 Global R&D funding forecast*. Washington, DC: Industrial Research Institute.

Khodr, H. (2011). The dynamics of international education in Qatar: exploring the policy drivers behind the development of Education City. *Journal of Emerging Trends in Educational Research and Policy Studies, 2*(6), 514–525.

Küster, A. (2010). Qatar—a new beginning? The development of a collaborative research and education environment in Education City, Doha. *Serials, 23*(1), 16–19.

MDPS. (2015). *Annual abstract 2015—chapter iv education statistics*. Doha, QA: Ministry of Development Planning and Statistics.

Mitchell, J. S. (2014). Beyond bricks and mortar: creating knowledge through student-faculty partnerships. *Journal of General Education, 63*(2–3), 73–93.

Powell, J. J. W. (2014). University roots and branches between "glocalization" and "mondialisation": Qatar's (inter)national universities. In A. W. Wiseman, N. H. Alromi, & S. Alshumrani (Eds.), *Education for a knowledge society in Arabian Gulf countries* (pp. 253–276). Bingley, UK: Emerald.

QF. (2013). *Qatar foundation strategic plan 2013–2023*. Doha, QA: Qatar Foundation.

QF. (2015). *Annual report 2013/14*. Doha, QA: Qatar Foundation.

QU. (2014). *Factbook 2013–2014*. Doha, QA: Qatar University.

QU. (2016a). *Academic programs*. Retrieved from www.qu.edu.qa.

QU. (2016b). External funding. Retrieved from www.qu.edu.qa.

QU. (2016c). *Undergraduate program requirements (Available for admission fall 2014)*. Retrieved from www.qu.edu.qa.

Salama, A. M., & Wiedmann, F. (2013). *Demystifying Doha*. Farnham, UK: Ashgate.

Shaaban, K., & Radwan, E. (2014). Rebuilding the transportation system in the city of Doha. *Journal of Traffic and Logistics Engineering, 2*(3), 241–247.

Sharifian, F. (2009). *English as an international language: perspectives and pedagogical issues*. Bristol, UK: Multilingual Matters.

The Peninsula. (2015). Qatar's R&D allocation hits QR3.25bn. *The Peninsula*. Retrieved June 18 from www.thepeninsulaqatar.com.

Vora, N. (2014). Between global citizenship and Qatarization: Negotiating Qatars new knowledge economy within American branch campuses. *Ethnic and Racial Studies, 37*(12), 2243–2260.

Wilkins, S. (2011). Who benefits from foreign universities in the Arab Gulf states? *Australian Universities' Review, 53*(1), 73–83.

Chapter 10
Students and Graduates

Abstract Students and especially graduates make up the workforce of tomorrow. Since Qatar is still in the early stages of its major transition which will go on for some or many more years, the current student body will play a major role in establishing a knowledge society. In Qatar, there are differences between Qatari and expatriate students, female and male students and also between public and private universities.

Keywords Qatar · Higher education · Universities · Students · Graduates Qatari students · Expatriate students · Females · Males · Public universities Private universities · Branch campuses · Education City · Qatar University

With education being one of the four important pillars of Qatar's National Vision, the country tries to improve its education system, pursuing two different strategies simultaneously: On the one hand, Qatar has launched a plan to modify its national university and on the other hand, the country imports renowned foreign universities and borrows their faculty and staff. The scholars and researchers of the branch campuses are faced with global principles while those of the national university are educated in the light of local traditions and norms like gender segregation (Moini, Bikson, Neu, & DeSisto, 2009; Powell, 2014). Qatar has identified several issues that still need to be addressed for the improvement of education and its outcome. The country tries, for instance, to increase motivation and awareness of the advantages that education entails, seeing the need to raise the number of students, especially that of males, in universities and other higher education institutions. Additionally, they try to overcome the tradition of school fees influencing the choice of the higher education institution a Qatari student is going to attend (GSDP, 2011).

J. Gremm et al., *Transitioning Towards a Knowledge Society*,
https://doi.org/10.1007/978-3-319-71195-9_10

10.1 Qatari Versus Expatriate Students

When it comes to the number of Qatari students as opposed to that of expat students, there is a clear difference between Qatar University and the private universities in the country. The majority, in fact about 66% of the students of Qatar University are Qataris; however, at Qatar's private universities the amount of Qataris and expats is more or less balanced. Despite the high number of expats, there are many Qataris (about 51%) studying at Education City's private universities (Interview partners 4 & 9). In Northwestern University for example, 43.3% of the whole student body in the academic year 2013/2014 are Qataris (NWU, 2014). Private colleges and universities in Qatar include the institutions of Qatar Foundation (mainly in Education City), North Atlantic College, Stenden University Qatar, Calgary University Qatar and Qatar College of Aeronautics (MDPS, 2015). Thereof Qatar Foundations' institutions have all in all about 2700 students and the College of the North Atlantic (a technical college operated by Canadian instructors) attracks about 3000 students (Table 10.1).

Only 48.6% of the total amount of enrolled students in private higher education institutions were expats. However, the number of graduates in Qatar (Table 10.2) in the academic year 2014/2015 did not correspond to the number of students in that year.

While the number of expat students was lower than that of Qatari students of both public and private universities in the academic year 2014/2015, the percentage of expat graduates studying in the private institutions is with 55.6% higher than that of Qataris. This is in line with Weber (2014) who mentions that the dominance of Qataris in numbers has changed and that the majority of Education City's graduates

Table 10.1 Number of enrolled students in Qatar by gender, nationality and sponsorship of the institution (2014/2015)

Sponsorship	Total			Qatari			Non-Qatari		
	Total	Female	Male	Total	Female	Male	Total	Female	Male
Public	**21,129**	15,231	5898	**13,919**	10,708	3211	**7210**	4523	2687
Private	**6977**	3316	3661	**3584**	1614	1970	**3393**	1702	1691
Total	**28,106**	**18,547**	**9559**	**17,503**	**12,322**	**5181**	**10,603**	**6225**	**4378**

Source MDPS, 2015

Table 10.2 Number of graduates in Qatar by gender, nationality and sponsorship of the institution (2014/2015)

Sponsorship	Total			Qatari			Non-Qatari		
	Total	Female	Male	Total	Female	Male	Total	Female	Male
Public	**1441**	1052	389	**793**	639	154	**648**	413	235
Private	**1793**	831	962	**796**	361	435	**997**	470	527
Total	**3234**	**1883**	**1351**	**1589**	**1000**	**589**	**1645**	**883**	**762**

Source MDPS, 2015

are expatriates. For expatriates working in Qatar it is difficult to find a university where they want to enroll their children. Many expat children are thus sent overseas to study (Interview partners 1 & 8).

One of the reasons why non-Qatari students are often enrolled in universities abroad is Qatar's sponsorship system. Expat students in Qatar face problems because of the *kafāla* system and identical requirements hold for them as for their parents. Students do not have their own sponsor, but are sponsored by one of their parents and, therefore, depend on the same *kafāla* as them. If the parent in question loses his or her sponsor, he/she needs to leave Qatar and the children have to leave with him or her, even if they are in the middle of their study (Interview partners 1, 4, 8, & 9). However, according to interviewee 10, it is possible for students to request a student visa from Qatar Foundation, so that they can stay in the country if the parent loses his or her job. At Qatar University, there are similar regulations. If the parent has to leave the country or the student loses his sponsor somehow else, the student may apply for a Qatar University sponsorship to carry on studying there. If people come to Qatar with the intention to study in the country, they need to apply for and fulfill the requirements to be granted a student visa, except for those who are granted a QU academic scholarship (QU, 2016a). People with a student visa or a scholarship of one of Qatar's universities need a sponsor as soon as they have finished their studies and want to start a job in Qatar. This does not hold for people coming from one of the other GCC countries. Those are allowed to study and work in Qatar without having a sponsor (Interview partner 1).

10.2 Female Versus Male Students

At the universities in Qatar, there is a strong imbalance of male and female students. Qatar has one of "the highest female-to-male university enrollment ratio worldwide" (Al-Kitbi, 2010, p. 98). All credit for the very good standing of women in Qatar's higher education institutions belongs to Sheikha Moza. "Sheikha Moza, the driving force behind Education City, is widely admired in the Arab-speaking world as a champion of women's rights and women's education" (Weber, 2014, p. 74).

Of all students enrolled in one of Qatar's universities in the academic year 2014/2015, 66.0% were female (Table 10.1). However, this excess of women did not hold when regarding only the private universities in Qatar. In the academic year 2014/2015, there were still a few more male students than female students enrolled in Qatar's private higher education institutions. The male students made up 52.5% of the student body. In the academic year 2014/2015, 58.2% of all Qatari graduates were females (Table 10.2). It is noticeable that a total of 73.0% of the graduates of Qatar's public universities were females, but that the number of male graduates superseded the number of female graduates in the private universities and higher education institutions of Qatar (male graduates: 53.7%). According to our interviewees, the number of female students in private universities has increased and has surpassed the number of male students. In Northwestern University, for instance, men account for a minority (Interview partner 4). That the ratio of males and females has changed in the private universities in the last few years, might be due to the fact that females have entered new fields like engineering and science only recently. Traditionally, they are advised to study in disciplines like education and health care, but Qatar has consciously increased the number of opportunities for females (Abouammoh, 2010; Al-Kitbi, 2010). Our interview partners mention a few guesses about why this imbalance of female and male students might exist: Interview partner 8 says that it is no secret that everywhere in the world women are more serious about studying. Furthermore, in Qatar, females work much harder than male students as they have the incentives to study. Since the future depends strongly on heritage and connections in Qatar, males are not as incited as girls (Interview partner 1). Abouammoh (2010, p. 84) states that "male students have a better chance to be employed in the industry with high school certificates". The author explains further that males are often interested in going to the military while girls usually do not enroll in military colleges and institutes. Another reason for men being outnumbered by women in Qatar's universities is that many male students leave Qatar to study overseas and therefore, there are fewer men than women in the higher education institutions of the country (Interview partners 1, 5, & 6).

Qataris do not really avoid studying in Qatar. It is rather that they prefer to study outside of the country if possible. "However, since international branch campuses have been established in the Gulf states, more nationals have decided to study at home" (Wilkins, 2011, p. 77). The large majority of Qataris who study somewhere else in the world are males (Interview partners 1 & 6). This is not due to women being neglected when trying to win a funded place or enroll at some degree programs that includes studying abroad. Females do not get the permission from their families to study away from home because traveling without accompaniment is not allowed by the parents (Davidson, 2010). Frank (2006, p. 47) reports, "because many girls are not allowed to go abroad alone to study, their best option for a first-class international education is Education City". From fifteen Qatari students that travel overseas to take their master classes, only one is female, while the rest are male. In the country itself, it is the other way around. While only two to three males try to get their master's degree in Qatar, twelve to thirteen females stay home to study for their master's degree. The fact that parents refrain from sending their daughters out of the country to study is not supported by the Emir of Qatar. He thinks that it is important for girls to study overseas and to come back to change the society, which is, accordingly, easier done by females than by males (Interview partner 10). Interviewee 6 reveals that the reason why academic staff is brought from other countries of the world is to teach the girls. This way, female students are offered the chance to receive high qualifications at a comparable international level (Lux, 2014). Qatari families happily send their sons away for education. They have enough money to serve their sons a good time overseas. There are many different motives for these stays abroad. One reason for studying abroad is the opportunity to acquire experience and to get to know the world and different cultures and traditions without being controlled by the parents.

Sometimes, however, studying abroad would not have been the initial choice of the students. Some of the males leave their environment because the programs they would like to study are not available in Qatar. Some other students are forced to enroll in universities outside of the country because their grades are not good enough to be accepted at one of the local universities (Interview partners 1, 5, 8, & 9). Traditionally, however, those who leave the country to get their degree overseas are the brightest students. Because of that, it is difficult to build a talented male student body at the local universities (Powell, 2014).

The motives of males and females to study are very different. Many of the students are eager to find good jobs and earn high salary after graduation. However, most of the Qatari women do not study to enter a good job. One reason for them seeking higher education is to improve their social prestige to marry some highly esteemed male. Men do not need to upgrade their prestige through education and might therefore not be as motivated (Interview partner 4; Fig. 10.1).

However, this is not the only motive for females to study. Some of the studying women in Qatar are already married and have children. Whether these women enter a job after graduation or not depends on their husbands because they have to finance their women completely. Most of the students in Qatar study because they really want to learn about the chosen subject (Interview partners 6, 8, & 9).

Fig. 10.1 The country encourages the people in Education City to try to achieve in studying, researching and inventing

10.3 Public Versus Private Universities

In the private universities of Education City, the enrolled students have to pay high tuition fees for their education. In these branch universities, Qatari and expat students pay the same. The tuition fees of Education City's branch campuses are adjusted to those at the home campus in the United States or Great Britain (Interview partners 1, 8, & 9). This is why the single tuitions vary between the different branch campuses. At Virginia Commonwealth University in Qatar the tuition amounts to 24,317 U.S. dollar per academic year (VCU, 2016) while students from Carnegie Mellon University in Qatar have to pay 51,196 U.S. dollar per year (CMU, 2016). In the private universities in the country, there are possibilities for students to apply for a scholarship and also many of the students get sponsored by companies in Qatar, so that they do not have to pay for their education. In the public university of Qatar, there are other regulations concerning tuition fees. While expat students have to pay there as well, Qatari students enjoy free education (Interview partners 1, 4, 5, & 6). For instance, expat students enrolled at Qatar University's College of Medicine have to pay about 25,000 U.S. dollar on average per year (QU, 2016b). However, according to Moini et al. (2009, p. 5), only "Qatari nationals who perform sufficiently well on national standardized school-leaving examinations are entitled to free higher education at the national university". Furthermore, the authors inform that Qatar University offers scholarships, however, only to some non-Qataris, given that Qataris do not have to pay to attend in the first place.

A lot of students in Education City study on a scholarship. In some universities, even about 50% of the whole student body receives financial aid. The sponsors are, for instance, government companies like Qatar Petroleum or Qatar Airways (Interview partners 1 & 5). At Carnegie Mellon University in Qatar, there are six different ways for students to receive financial aid. Scholarships exist from the Qatar Supreme Education Council, from Sheikh Nasser bin Hamad bin Abdullah Al Thani, from Qatar Foundation and from Carnegie Mellon University. Local companies provide sponsorships as well and Qatar Foundation additionally offers a loan to some students. As an example: Out of 427 enrolled students, 354 qualify for these scholarships (CMU, 2015). Nevertheless, companies do not grant scholarships to students without a service of the students in return. Students agree on working in the company that sponsors them for a few years after graduation. Some of the students already work in the companies, for instance as interns or student associate during studying. This kind of agreement between a student and a company is problematic if the subject of the degree course differs from the profession of the company. If this is the case, students are not educated properly to serve the company adequately. This often results in dissatisfaction on both sides (Interview partners 1, 6, & 9).

Beside the different regulations regarding tuition fees in Qatar's private and public universities, there are different ways of dealing with education of female and male students. Qatar University offers gender-segregated bachelor degree courses. Every course exists two times, but one of them is only for male students and the other only for female students (Interview partners 1, 4, 5, 6, & 8). However, men and women are not only separated during the courses, there are also separate campuses and libraries for female and male students at Qatar University (Weber, 2014). This gender separation only holds for undergraduate students, though. The same faculty that teaches male students also teaches female students (Moini et al., 2009). The segregation of male and female students was introduced for females to participate in classes while still respecting traditional family values. However, this separation does not hold for post-graduate studies where females and males attend the same courses (Davidson, 2010; Interview partner 10). Moreover, there is no gender segregation when it comes to events hosted by Qatar University. As opposed to this, students at private universities visit the same courses and share the same campus regardless of gender (Interview partners 4 & 10; Fig. 10.2).

Since the academic year 2007/2008 all branch campuses in Qatar teach in a coeducational environment. Before that, the studies in Virginia Commonwealth University were only open for women (Moini et al., 2009). Sheikha Moza is strictly against the separation of female and male students in Education City. Only the student dorms are gender-segregated there. Parents who send their children to one of the private universities in Qatar know that males and females are not trained separately and accept the concept of coeducation (Interview partners 4, 10, & 11). According to interviewee 1, the only advantage of gender-segregated studies is that the parents of the females are happy. Apart from that he can only think of

Fig. 10.2 Mixed-gender hall at Carnegie Mellon University in Qatar

disadvantages: Since females and males are mixed in companies later in life, they should learn how to work together as early as possible. Especially "skilled professionals will increasingly be required to function in gender-mixed environments" (Moini et al., 2009, p. 31). Education City's branch campuses offer the chance to do that and prepare female and male students for this situation.

Qatar's public and private universities neither see the same problem, nor take the same measures, when it comes to problems with English as the teaching language. In countries, where English is not the first spoken language, there always occur problems with English as the teaching language. This also holds for the universities in Qatar (Interview partner 6). Especially in Qatar University, where most of the students are Qataris, the students have trouble speaking and writing in English fluently. Until 2005, teaching at QU was mainly in Arabic and switched to English since then. In 2013, Qatar University decided that the main teaching language should be Arabic again. So they switched it back and retained this as the main language till today (Interview partner 8), even though speaking English is a requirement for working in the private and public companies in Qatar. This backward motion was caused by the fear of the local people that the younger generations would neglect Qatar's national identity. Students and academics were also of the opinion that this would allow more students to enroll at QU because they would not need to pass any tests concerning their English skills any longer (Haroon, 2012). In the universities in Education City, there are less problems with English as teaching language. All of the students need to pass the test of English as a Foreign Language before they start their studies (Interview partner 2). However, this measure does not entirely eliminate the problems.

> Despite application requirements intended to ensure a high proficiency in English, […] an alarming number of local students demonstrate remarkably weak skills when it comes to producing, evaluating, understanding and synthesizing exactly the sort of complex texts they will encounter in knowledge-intensive occupations (Ulmer, 2009, p. 110).

Another step to reduce the issue of students having trouble with the English language was to keep the number of students, especially that of Qatari students, low to give the academic staff more time to supervise each student individually (Interview partner 10). The average class size in Carnegie Mellon University is, for instance, 19 students (CMU, 2015). "Qatar Foundation has also established an Academic Bridge Program to provide supplementary preparation—in English […]—for Qatari secondary school graduates who aspire to enroll in one of the Education City branch campuses" (Moini et al., 2009, p. 7).

10.4 Summarizing Qatar's Higher Education Conditions

Qatar's students and graduates are the knowledge workers of tomorrow who will be responsible for the changes in the country's society. In the state of Qatar, there are strong differences between Qatari versus expatriate students and graduates as well as female versus male students and graduates. While Qataris prefer to go to public universities, expatriates favor studying at the private universities of Qatar. Since only the best of Qatari secondary school graduates are entitled to enjoy the high-quality education provided by the branch campuses in Qatar, those Qatari nationals who do not qualify, enroll at Qatar University (Moini et al., 2009). There, they are educated in Arabic and in a gender-segregated environment. This might become a disadvantage when they change over to the labor market, where the fluency in English is necessary and where men and women are required to work with each other professionally. In contrast to the students of Qatar University, the students of the private universities are prepared for these circumstances and, therefore, might be more successful in their occupations. The number of female students is higher than that of male students. While this might be partially caused by the fact that women are not allowed to study abroad, it cannot be denied that females are more ambitious when it comes to studying. The lack of interest of male students is problematic. If the number of those students decreases further, there will soon be an insufficient amount of well-educated male graduates for companies to recruit. Females are then the ones who might take over the professional and managerial jobs in a society that is traditionally male dominated (Kosior, Barth, Gremm, Mainka, & Stock, 2015).

One of the most crucial parts in the life of a student is graduation and the subsequent transition into the labor market. During studying, students acquire the needed knowledge which they can finally apply in their first job after graduation. This is where the country really benefits from great and outstanding graduated students. What about the graduates in Qatar? Do they face the same conditions regardless of gender and nationality when searching for a job? Does the motivation of graduates differ? What are the preferred workplaces for female, male, Qatari and expat graduates?

References

Abouammoh, A. M. (2010). The role of education: reform trends and EU-GCC cooperation. In C. Koch & L. Stenberg (Eds.), *The EU and the GCC: challenges and prospects under the Swedish EU presidency* (pp. 77–94). Dubai, AE: Gulf Research Center.

Al-Kitbi, E. (2010). Where do GCC women stand in the development process? In C. Koch & L. Stenberg (Eds.), *The EU and the GCC: challenges and prospects under the Swedish EU Presidency* (pp. 95–103). Dubai, AE: Gulf Research Center.

CMU. (2015). *Carnegie Mellon University Qatar: annual report 2014–2015*. Doha, QA: Carnegie Mellon University.

CMU. (2016). *Tuition and costs of attendance*. Retrieved from www.qatar.cmu.edu.

Davidson, C. M. (2010). The higher education sector in the Gulf: history, pathologies, and progress. In C. Koch & L. Stenberg (Eds.), *The EU and the GCC: challenges and prospects under the Swedish EU Presidency* (pp. 59–75). Dubai, AE: Gulf Research Center.

Frank, L. (2006). Qatar taps wells of knowledge. *Science, 312*(5770), 46–47.

GSDP. (2011). *Qatar national development strategy 2011–2016*. Doha, QA: General Secreteriat for Development and Planning.

Haroon, A. (2012). Elation, worries on campus. *The Peninsula*. Retrieved January 27 from www.thepeninsulaqatar.com.

Kosior, A., Barth, J., Gremm, J., Mainka, A., & Stock, W. G. (2015). Imported expertise in world-class knowledge infrastructures: the problematic development of knowledge cities in the Gulf region. *Journal of Information Science Theory and Practice, 3*(3), 17–44.

Lux, C. (2014). Qatar national library: architecture as innovation in the Arab world. *International Federation of Library Associations and Institutions, 40*(3), 174–181.

MDPS. (2015). *Annual abstract 2015—chapter iv education statistics*. Doha, QA: Ministry of Development Planning and Statistics.

Moini, J. S., Bikson, T. K., Neu, C. R., & DeSisto, L. (2009). *The reform of Qatar University*. Santa Monica, CA: RAND Corporation.

NWU. (2014). *Envisioning the future: year in review 2013–2014*. Doha, QA: Northwestern University.

Powell, J. J. (2014). University roots and branches between "Glocalisation" and "Mondialisation": Qatar's (inter)national universities. In A. W. Wiseman, N. H. Alromi, & S. Alshumrani (Eds.), *Education for a knowledge society in Arabian Gulf countries. International perspectives on education and society* (pp. 253–276). Bingley, UK: Emerald.

QU. (2016a). *Students visa*. Retrieved from www.qu.edu.qa.

QU. (2016b). *Undergraduate admissions—tuition fees*. Retrieved from www.qu.edu.qa.

Ulmer, J. (2009). The knowledge society: text and orality in Qatar. *The International Journal of the Book, 6*(1), 109–113.

VCU. (2016). *Costs of (BA/BFA) study*. Retrieved from www.qatar.vcu.edu.

Weber, A. S. (2014). Education, development and sustainability in Qatar: a case study of economic and knowledge transformation in the Arabian Gulf. In A. W. Wiseman, N. H. Alromi, & S. Alshumrani (Eds.), *Education for a knowledge society in Arabian Gulf countries. International perspectives on education and society* (pp. 59–82). Bingley, UK: Emerald.

Wilkins, S. (2011). Who benefits from foreign universities in the Arab Gulf states? *Australian Universities' Review, 53*(1), 73–83.

Chapter 11
Transition of Graduates into the Labor Market

Abstract The labor market situation in Qatar is, as well as in the GCC states, quite different from the situation in European countries. The market is overcrowded by foreign people—workers and talents. Qatar is confronted with this fact and has to decide how to handle it in the future. A good basis is provided by excellent study conditions. But a second key aspect—the motivation of Qataris to find a job in the private sector—leaves much to be desired. They rather rest in the public sector, where the working conditions and the payment are excellent. First approaches to change the situation have made only minor results until now and there is yet no end in sight.

Keywords Qatar · Higher education · Universities · Graduates
Labor markets · Qataris · Expats · Post-graduates · Workforce
Private sector · Public sector · Job opportunities · Employment
Qualification · Motivation · Females · Males · Qatarization

The main goal behind the development of high-class educational facilities is to prepare students to work not only in the national economy, but rather in a global, internationally acting world (Khodr, 2011). The purpose of retaining graduates within the own workforce is more than important in Qatar (QF, 2013). However, to survive in a cross-national environment, it is crucial to be able to think globally—that holds true for both company and workforce. By training the relevant topics with the help of excellent resources and teachers, it is possible to shape a group of "graduates with international perspectives in a faster and certainly more effective and efficient way" (Khodr, 2011, p. 522). The multinational way of education—realized in Education City as an inviting place for international universities and in Qatar University as home of international faculty, students and cooperation—becomes imperative with this goal and tries to enable the creation of "a research culture and capacity in the country to support the knowledge-based economy" (Ibnouf, Dou, & Knight, 2014, p. 46). The development of new knowledge and highly competent job starters is supported by different initiatives like internships or the Qatar Science Leadership Program that shall help the young generation to

J. Gremm et al., *Transitioning Towards a Knowledge Society*,
https://doi.org/10.1007/978-3-319-71195-9_11

develop leadership skills and to find jobs even in higher positions. In particular, the consequential capable, motivated workforce is able to build up “a strong public sector” and “a thriving economy, particularly one that is ‘knowledge-based’” to “deliver sustainable development and quality of life for its people” (QF, 2013, p. 16). As the general basis, the skills and the potential of the nation are both repeatedly mentioned. The population is thus called to become more active, to come to appreciate lifelong learning and career achievement and to develop an interest in research (QF, 2013). Only if all these plans are put into practice, people will get started and the training will show success, the programs of nationalizing the workforce could be successful. Currently, it is true that Qatar “has a limited pool of engineers and limited outlets for producing engineering graduates” (Abdulwahed, Pokharel, Balid, & Hasna, 2013).

11.1 Level of Graduates: Qataris Versus Expats

It is already recognized by QF and discussed in their Strategic Plan 2013–2023 that Qatar’s students are at the K-12 level not very well-educated. Year 9 students in Qatar are on average three years behind those of other OECD members and even five years behind students from cities like Singapore that are still more educationally advanced. The problem is identified and shall be eliminated by significantly improving K-12 education outcomes and refining the universities’ curricula, like for example Texas A&M University in Qatar does, so that both go well together with Qatar’s, the region’s and the world’s economic needs (Parsaei & Elbashir, 2014; QF, 2013). Otherwise, and as currently is the case, after high school, students are not “well prepared for tertiary options and the job market” (QF, 2013, p. 16).

After attending and graduating from one of Qatar’s public or private universities, it seems to be entirely different. Graduates are adequately educated, they are able to meet their challenges on the labor market and are flexible in choosing what they would like to do next (Interview partners 5 & 6).

The compensation of these differences after high school graduation is supported by different Foundation Years or Academic Bridge Programs. They shall help to prepare the underprepared Qataris and expatriates so that they are able to make their mark (Weber, 2014). To measure what the students do well during their study and what could be improved regarding both the students’ behavior and the training staff’s teaching method, Qatar University even implemented a learning outcomes assessment system (Al-Thani, Abdelmoneim, Daoud, Cherif, & Moukarzel, 2014). Interview partner 1 assures that every Education City graduate has an employment after seven to twelve months resulting from the fact that most of them have a sponsor waiting for the graduate. The others confirm that this holds for the Qataris and that they do not have any problems to find employment. They have the choice between different job opportunities and get multiple offers (Interview partners 6 & 12). It does not depend on the degree or the kind of graduation, it is just a matter of fact. Despite the fact that they are “on average better educated than foreign workers"—not experts—"they are still not sufficiently educated to fill the positions of highly skilled labor" (Berrebi, Martorell, & Tanner, 2009, p. 435).

The situation is a little different regarding the expatriate graduates. Expat undergraduates, especially women, do not find a job, however, after graduation, it looks better. In terms of Qatarization, many companies prefer to employ Qataris—in some cases they are even bound to hire nationals. If, however, others are better educated, why should they choose them (Interview partners 1 & 8)? That way of thinking seems to be the chance for the young educated expats. However, the import of experts that advance the country's development with their experience can get in their way so that expats with zero experience have again huge difficulties to find an employment. Starting by working in a side job in the field of the desired job or the specialization in a field where not that many people are skilled, are certainly the only chances for the young foreign elite in Qatar (Interview partner 12). Otherwise and if they do not find a sponsor, the young talents have to leave the country. This is why many expats start to move away now. They are worried about their children's future due to existence problems coming up with the decreasing oil prices (Interview partner 4). The graduates' parents came to the country because they were needed, but that does not mean that the graduates themselves are needed as well. In some cases, expats are employed just for one project, like a foreign librarian who was asked to build up the new national library, but after finishing the buildup the job will be over and another person will take over the position. This is the way how Qatar's knowledge society development works (Interview partner 12).

Whether Qatari or expat graduates are better off depends on the market and the discipline, but in each and every case connections are important. Your own reputation can get you a long way regardless of your origin (Interview partners 6 & 12). Based on our knowledge, it is difficult to figure out if there are many opportunities for the graduates to find work in a knowledge-intensive sector—both public and private—because our interview partners did not have any information about it and also the literature does not provide any data.

11.2 Different Paths Taken by Post-graduates

Many students—both Qataris and expats—choose for different reasons different post-graduate paths. Some Qataris do not even search for a job and rather go abroad to London or the like for an internship because their families have residences there. Others take this opportunity already during their study. Men usually prefer this opportunity due to moving freely in the foreign country, which is not possible in their home country, and possibly study especially for this reason (Interview partner 4). If women choose a scientific career, it is in most cases over as soon as they get married. Despite the high number of female students "only a small percentage of these graduates ever enter the workforce" (Coffman, 2005, p. 256). They have the same opportunity to leave the country, but not alone (Interview partner 13). "A female student going for postgraduate studies abroad must be accompanied by a *mahrum* (a male relative serving as a chaperon)" (Bahgat, 1999, p. 134). However, most Qataris rather enjoy their cradled luxury life—enabled by the distribution of the country's revenues—which is reflected by the high number of unemployed under 25 year olds (more than 25%) in the GCC (Hoteit, Tuerpitz, & Luelsdorf, 2015). Then again, others (mostly expats) are interested in lifelong learning and education after graduation—they see updating knowledge as a kind of investment and decide for continuing education (Interview partners 4, 5, & 6). Otherwise, they probably have no chance to survive on the nation's labor market that is again and again refilled with highly educated foreign talents.

Like probably in most countries, women have major difficulties to find leadership positions. In general, "women graduates only enter a limited number of professions, such as teaching, where they are not required to interact with unrelated males" (Weber, 2014, p. 73). Despite this cultural context, in most cases, it rather depends on the social class and the family (Interview partners 5 & 8). However, exceptions confirm the rule. The former president of Qatar University was Sheikha Abdulla Al Missned (QU, 2016c), the director of the National Museum of Qatar is Sheikha Amna bint Abdulaziz bin Jassim Al Thani (Qatar Museums, 2016b) and Sheikha bint Ahmed Al Mahmoud is the minister of education (Al Baik, 2003). In addition to it, lots of females generally work in libraries, and some sectors, like the health care sector, include positions exclusively for women. In many cases, the marriage blocks the women's career, but there are also some exceptions of women that are married and study at the same time, that marry first and work thereafter or do both at the same time (Interview partners 1, 5, & 12). To raise the number of Qatari workers, students and graduates are getting promoted and especially the highly motivated women get encouraged in the male-dominated society to reduce the women's difficulties in finding jobs in leadership positions (Al-Kitbi, 2010; Interview partner 12). The government even implemented a program that shall "develop skills in the workplace, such as various executive education programs" (QF, 2013, p. 22) to strengthen the qualification of the graduates. The program is called the Qatar Science Leadership Program (QSLP). Considered as a whole, when looking at those graduates who aim at lower positions, there are no differences between men and women (Interview partner 1). However, when it comes to family and tribal relations, problems may arise for women. "As Qatari women attend and graduate from institutions of higher education and many enter the work force, their mobility and visibility increasingly juxtaposes their roles in the family and tribe with their new roles as partners in the creation of a nation" (Asmi, 2016, p. 18).

11.3 Private Versus Public Sector as Employer Prospect

Far from the uneven distribution of job opportunities for natives and expats as well as males and females, the preferences of choosing a working sector seem to be an interesting labor market topic. Since most Qataris get a job anyway, earn about 60% more than the expats after receiving their undergraduate degree and choose the job by comparing benefits, only few search for self-actualization on their job. They prefer the public sector in almost all cases because the work is less stressful, less challenging, better payed and the working times are more comfortable (Interview partners 1, 5, 8, & 12). In addition, the salary for Qataris in the public sector is up to three times higher than in the private (Randeree, 2012) because "employment in the civil service is used as one of the key ways for the government to distribute the nation's oil wealth among Qatari citizens" (Berrebi, Martorell, & Tanner, 2009, p. 435). As the Qataris currently studying are the next generation, one has to query whether it is a good idea to rely on this rather unmotivated group of people.

In contrast, expats are more interested in the economic opportunities and are motivated to perform well, which can be better realized in private companies. In general, they are willing to work more hours, agree to lower salaries, accept poorer working conditions and have no problem to do more physically demanding jobs than any wealthy Qatari (McMurray, 1999). They have more difficulties to get a sophisticated, well-paid job and try to win over the employers with the prospect of keeping pace with imported expertise of expats from the outside who, in turn, come because of the good salary. In the public sector, their competitors are the preferred Qataris (Interview partners 1 & 6; Fig. 11.1).

This distribution of a—compared to all other sectors—high number of Qataris in the public sector and a private sector full of expats is shown in Table 11.1.

The preferences are unambiguous, even if no sector is reigned exclusively by Qataris because non-Qataris interfere in every field (even the public sector is to over 50% in the expats' hand). About 76,000 of overall approximately 94,000 Qataris (in detail 81.1% of the whole Qatari workforce) work in government departments or

Fig. 11.1 Qatar's West Bay area houses different companies that are potential employers of Qatar's graduates

Table 11.1 Economically active population by nationality and sector (2014)

Sector	Total number of employees	Qatari (% of all Qataris)	Non-Qatari (% of all non-Qataris)
Government department	140,672	64,343 (68.79)	76,329 (4.79)
Government company/ corporation	62,051	11,483 (12.28)	50,568 (3.17)
Mixed	55,048	6364 (6.80)	48,684 (3.06)
Private	1,270,913	10,830 (11.58)	1,260,083 (79.08)
Diplomatic	2681	91 (0.10)	2590 (0.16)
Non profit	1439	421 (0.45)	1018 (0.06)
Domestic	154,057	0 (0.00)	154,057 (9.67)
Total	1,686,861	93,532 (100.00)	1,593,329 (100.00)

Source Qatar Information Exchange (2014)

government companies and/or corporations. In all other sectors there works between 6.8 and 0.0% of the Qatari workforce. Looking at the non-Qataris, the distribution is different: 79.1% of the whole non-Qatari workforce work in private sectors. However, it results in just about 8.0% of the expats working in a government department or a government company or corporation. In contrast, 11.6% of the employed Qataris are active in the fully privatized sector. These findings in turn support clearly the previous result that Qataris prefer the public sector because of good conditions, although they probably would get a job wherever and whenever they wanted. Expats, in contrast, prefer—more or less inevitable, but exactly fitted to their purposes—the private sector. Taken as a whole, it seems to be a fact that "females are not looking so much for financial incentives but rather for benefits" and on the contrary "males need large financial packages" (Interview partner 12).

Overall, the number of vacant jobs becomes less these days because of the overcrowded workforce. Two years ago, it was completely different and many jobs were offered (Interview partners 5 & 6). A sector in which graduates are still needed is the public museums sector. Owing to the construction of the National Museum of Qatar that houses, among others, galleries, a research center and laboratories, many museum-specific jobs are generated (Qatar Museums, 2016a; Interview partner 13).

11.4 Can an Increase in Motivation Promote Qatarization?

Qatarization "refers to the government's efforts to create sufficient employment opportunities for nationals and to limit the dependence on expatriate labour" (Al Muftah, 2016, p. 283). To raise the motivation for entering the workforce in a successful and ambitious way and to bring the Qataris also into the private sector,

Qatar's universities cooperate with different companies like Vodafone, General Electric Power or some institutions from the healthcare sector. Thus, students shall gather hands-on experience regarding their future work life. The companies providing these internships are not necessarily located in Qatar. For instance, the Northwestern University in Qatar affords the opportunity of collecting experience during a journalism program outside of Qatar for two months. In some cases, even women participate in these types of programs. Several courses integrated internships as a regular mandatory feature (Interview partners 1, 5, 6, & 9). It has not been pushed enough, yet—too few meaningful career services and work experiences are adequate—because there are still enough jobs, too much money and less pressure (Hoteit, Tuerpitz, & Luelsdorf, 2015; Interview partner 12). However, the main goal is to provoke the students or graduates and especially the Qataris to find broad interest in working in general and in working in the suitable company or sector (Interview partner 6).

The preferred and most needed economic fields for nationalizing the workforce are thereby science, technology, engineering and mathematics (STEM) that in turn claim the education of highly skilled and highly knowledgeable workers (Gonzalez, Karoly, Constant, Salem, & Goldman, 2008; Barber, Mourshed, & Whelan, 2007). With the help of motivated, well-educated teachers, whose expectations and relationships strongly influence the youth's decision for or against a career after graduation (Wiseman, Alromi, & Alshumrani, 2014), the nationals' lack of knowledge, skills and competencies can—in collaboration with those work experience programs—hopefully play a part in contributing to the reduction of the high unemployment rate of young citizens (Hoteit, Tuerpitz, & Luelsdorf, 2015).

The official program of nationalizing the private workforce goes one step further. It concentrates on all Qataris and not only on the young generation. These days, many important economic decisions are made by non-Qataris and therefore non-native experts have the power to influence the country and its inhabitants. Qataris want to break out of the dependence on the expats and make their own decisions (Interview partner 1). With being present in private firms, Qataris could "bring a deep connection to their nation's needs and values" (QF, 2013, p. 20). Until now, they are far away from reaching this point and from being independent. The Emir supports this nationalizing venture, trusts in the idea that it "will lead to higher national employment in high skill and high salary positions" (Wiseman, Alromi, & Alshumrani, 2014, p. 6) and has the slightly daring vision that Qataris that take the expats' workplaces will indeed work (Interview partner 10). Plans for a Qatarization of the workforce are thus existing and in some sectors successful, like the banking and airline industries, the field of museums and certainly governmental offices. If a qualified Qatari is found, he or she is hired and the expat discharged. Especially for women, one can identify a trend towards accepting jobs in private companies. While in 2001 only 2.1% of the female labor force was employed by a private company, in 2013 the relative frequency increased to 14.7% (Kassem & Al-Muftah, 2016, p. 223).

However, in the end no one believes in full Qatarization. Despite this attempt, expatriate talent will still be brought to the country (Interview partners 1, 5, 6, 8, & 13). Replacing the whole or at least a large part of Qatar's labor market will take some generations and is not realizable within just a few years. The planners think it is possible to recreate a prestigious university landscape like it is the case in Harvard or Cambridge, and that they will have Qatari graduates that are at the same level as students from those universities in the end, but it is not that easy as they expect. Without any opportunities and will to come together and enjoy the evening in pubs or similar places, a culture like the one we know from Western regions cannot emerge. Western work life with all its facets will be hard to implement, so that the country will probably draw again and again on foreigners if the nationals do not change their attitude. Furthermore, the fact that Qatar is a tribal society in which women do not transfer citizenship to their children, will impede the state's progress (Interview partners 1 & 4). In the country, "as a result of particular state-society relations, welfare benefits are exceedingly high for nationals, and consequently, the state shows great reluctance to expand citizenship, as it would be economically burdensome" (Babar, 2014, p. 403). Just around fifty foreign people including important people like doctors or handball players to fill the national team will be naturalized each year (Interview partner 4). As a result, a fusion of different cultures and personal attitudes is scarcely possible, which makes the conversion of the natives' motivation and working morale more and more difficult. From a formal point of view, Qatarization seems to be promising, but whether it is a good idea remains open.

11.5 Summarizing the Possibilities of Qatar's Graduates

Owing to a good education, the majority of graduates is adequately prepared to start a successful working life. Qataris choose—if they decide to work at all—the public sector, clearly because of the conditions. They love their luxury life and, therefore, only very few choose the harder way of working in a private company. Expats have a rougher ride. Despite harder conditions, they want to go into the private sector (maybe because of the fact that they would barely have any chances in the public one), compete there with Qataris or newly brought expats and have the incentive to win the challenge, but do not always succeed. The attempts to normalize the unusual situation have shown minor results until now, but the government follows up on its plans. The country seeks to do excellent trainings "to develop young and promising nationals" (QF, 2013, p. 29). However, it does not forget the expatriates that shall "contribute to the nation's development, by entering the Qatari workforce or by generating new knowledge that can support the economy" (QF, 2013, p. 29). Concerning the nationals, the majority of our discussion partners do not believe in full implementation of Qatarization. They would have to change their unmotivated attitude, but because they probably will not do this, the "transfer of knowledge from expatriate[s] to citizen[s]" (Randeree, 2012, p. 1) does not proceed and the citizens do not adapt to the labor market, so that Qatarization will not be successful. In contrast, expatriate graduates will suffer more and more, if the plan is someday completed. Despite the fact that they are included in the nation's plans, they are eliminated step-by-step by being the second choice in the selection of new employees, so that they are ambiguous about their future in Qatar. All told, the country's situation is ambivalent. Too many expats would restrict the nation's culture; too few would cause a lot of setbacks for the development of Qatar because the nationals alone cannot rescue the situation. There are two main issues summarized by Hvidt (2015, p. 45) the country has to deal with:

> [F]irst, they must place an increased emphasis on educating their own people to become knowledge workers and, second, they must be alert and adaptive in their policies relating to providing the expats workers with appropriate incentives to stay; incentives such as upholding salary levels and benefit packages that are competitive internationally.

Whether they will find a fair way to balance both aspects is not clear until now. And if they do, it will take a long time. However, "Qatar may not be changing as fast the world would like, but the slow and steady pace is being managed so that" (Khatri, 2015, 1st paragraph) Qatar's young population acquires the knowledge how to deal with it.

The topic of universities has been discussed in detail now, completed by a description of the higher education conditions for both students and graduates and of paths of the post-graduates. Beside the university as a knowledge-intensive institution, another important one may not be ignored: the library. Are libraries already developed in Qatar? To which extend are they able to contribute to the emergence of a knowledge society? Which other knowledge-intensive institutions could additionally assist in this purpose?

References

Abdulwahed, M., Pokharel, S., Balid, W., & Hasna, M. O. (2013). Systemic framework of competencies of 21st century technical and engineering human resource. A development case for a knowledge based economy in Qatar. In *7th IEEE GCC Conference and Exhibition, November 17-20, 2013, Doha, Qatar* (pp. 413–418). Washington, DC: IEEE.

Al Baik, D. (2003, May 7). Al Mahmoud is Qatar's first woman minister. *Gulf News*. Retrieved from www.gulfnews.com.

Al-Kitbi, E. (2010). Where do GCC women stand in the development process? In C. Koch & L. Stenberg (Eds.), *The EU and the GCC: Challenges and prospects under the Swedish EU presidency* (pp. 95–103). Dubai, AE: Gulf Research Center.

Al-Muftah, H. (2016). Demographic policies and human capital challenges. In M. E. Tok, L. R. M. Alkhater, & L. A. Pal (Eds.), *Policy-making in a transformative state. The case of Qatar* (pp. 271–294). London, UK: Palgrave Macmillan.

Al-Thani, S. J., Abdelmoneim, A., Daoud, K., Cherif, A., & Moukarzel, D. (2014). Assessment of student learning outcomes for assurance of learning at Qatar University. *Journal of Assessment and Institutional Effectiveness, 4*(2), 116–136.

Asmi, R. (2016). Finding a place to sit: How Qatari women combine cultural and kinship capital in the home majlis. *Anthropology of the Middle East, 11*(2), 18–38.

Babar, Z. R. (2014). The cost of belonging: Citizenship construction in the state of Qatar. *Middle East Journal, 68*(3), 403–420.
Bahgat, G. (1999). Education in the Gulf monarchies: Retrospect and prospect. *International Review of Education, 45*(2), 127–136.
Barber, M., Mourshed, M., & Whelan, F. (2007). Improving education in the Gulf. *The McKinsey Quarterly*, 39–47.
Berrebi, C., Martorell, F., & Tanner, J. C. (2009). Qatar's labor markets at a crucial crossroad. *Middle East Journal, 63*(3), 421–442.
Coffman, J. (2005). Higher education in the Gulf: Privatization and Americanization (Fall 2003). In P. G. Altbach & D. C. Levy (Eds.), *Private higher education: A global revolution* (pp. 255–260). Rotterdam, NL: Sense Publishers.
Gonzalez, G., Karoly, L. A., Constant, L., Salem, H., & Goldman, C. A. (2008). *Facing human capital challenges of the 21st century*. Doha, QA: RAND-Qatar Policy Institute.
Hoteit, L., Tuerpitz, A., & Luelsdorf, L. (2015). *Motivation and perseverance: The missing link to GCC knowledge economies*. Dubai, AE: The Boston Consulting Group.
Hvidt, M. (2015). The state and the knowledge economy in the Gulf: Structural and motivational challenges. *The Muslim World, 105*(1), 24–45.
Ibnouf, A., Dou, L., & Knight, J. (2014). The evolution of Qatar as an education hub: Moving to a knowledge-based economy. In J. Knight (Ed.), *International education Hubs: Student, talent, knowledge-innovation models* (pp. 43–61). Dordrecht, NL: Springer, Netherlands.
Kassem, L. M., & Al-Muftah, E. (2016). The Qatari family at the intersection of policies. In M. E. Tok, L. R. M. Alkhater, & L. A. Pal (Eds.), *Policy-making in a transformative state. The case of Qatar* (pp. 213–239). London, UK: Palgrave Macmillan.
Khatri, S. S. (2015, July 11). Sheikha Moza: World should be patient with Qatar. *Doha News*. Retrieved from www.dohanews.co.
Khodr, H. (2011). The dynamics of international education in Qatar: Exploring the policy drivers behind the development of Education City. *Journal of Emerging Trends in Educational Research and Policy Studies, 2*(6), 514–525.
McMurray, D. (1999). Recent trends in Middle Eastern migration, trafficking, and transiting: New perspectives on labor migration. *Middle East Report, 21*(1), 16–19.
Parsaei, H. R., & Elbashir, N. (2014). A new vision for engineering education. In *Proceedings of the 121st ASEE Annual Conference & Exposition, Indianapolis, IN, June 15–18, 2014* (24.80; 8 pp.). Washington, DC: American Society for Engineering Education.
Qatar Information Exchange. (2014). *Labor force sample survey 2014*. Doha, QA: Qatar Information Exchange.
Qatar Museums. (2016a). National museum of Qatar. Retrieved from www.qm.org.qa.
Qatar Museums. (2016b). Our senior leadership—shared passion. Retrieved from www.qm.org.qa.
QF. (2013). *Qatar Foundation Strategic Plan 2013 – 2023*. Doha, QA: Qatar Foundation.
QU. (2016c). History of the office. Retrieved from www.qu.edu.qa.
Randeree, K. (2012). *Workforce nationalization in the Gulf Cooperation Council States*. Doha, QA: Center for International and Regional Studies.
Weber, A. S. (2014). Education, development and sustainability in Qatar: A case study of economic and knowledge transformation in the Arabian Gulf. In A. W. Wiseman, N. H. Alromi, & S. Alshumrani (Eds.), *Education for a knowledge society in Arabian Gulf countries: International perspectives on education and society* (pp. 59–82). Bingley, UK: Emerald.
Wiseman, A. W., Alromi, N. H., & Alshumrani, S. (2014). Challenges to creating an Arabian Gulf knowledge economy. In A. W. Wiseman, N. H. Alromi, & S. Alshumrani (Eds.), *Education for a knowledge society in Arabian Gulf countries: International perspectives on education and society* (pp. 1–33). Emerald: Bingley, UK.

Chapter 12
Libraries, Science Parks and Research Funding

Abstract Institutions like libraries and science parks are knowledge-intensive institutions where knowledge is produced and maintained. Therefor their presence is important for a country's development. Qatar has realized the importance of these institutions and is currently trying to build a new national library in Education City. Also there is a science park in the same area. The proximity of knowledge-intensive institutions makes collaboration possible. This, together with the funding of research, might support the country on its way into the knowledge society.

Keywords Qatar · Libraries · Librarians · Qatar National Library
Media and information literacy · Science parks · Qatar Science and Technology Park · Sidra Medical and Research Center · Research funding · Qatar National Research Fund · Triple helix partnerships

"Higher education and research, as much as a knowledge-based economy in general, need access to international information and knowledge and the development of a national knowledge base" (Lux, 2014, p. 175). According to Mainka and Khveshchanka (2012), libraries are important knowledge-hubs of knowledge-based societies because they are the traditional institutions of knowledge management. Beside print publications, libraries need to provide sufficient digital material and attractive physical infrastructure for their users. Science and technology parks are important research institutions to produce and process this knowledge in the first place. The focus is on scientific, technical and medical (STM) knowledge, but everyday knowledge is not left out in these institutions (Stock, 2015).

The political willingness of the administration is crucial for the change towards a knowledge-based society. The leaders design a master plan and fund the planned transformations and innovations adequately by public or private funds (Stock, 2015). To support Qatar on its way from a carbon-based economy to a diversified and knowledge-based economy, the Qatar Foundation for Education, Science and Community Development has initiated some large projects, including the establishment of Qatar National Library, Qatar Science and Technology Park and Sidra Medical and Research Center whose impressive high-class buildings are all located

J. Gremm et al., *Transitioning Towards a Knowledge Society*,
https://doi.org/10.1007/978-3-319-71195-9_12

in or close to Qatar's Education City. The foundation of Qatar National Research Fund has also been initiated by Qatar Foundation. All of these initiatives support knowledge-based development in Qatar adhering straight to the ideas of the country's National Vision. Qatar National Library opened in November 2017. As national library, university and research library as well as central public library, it provides knowledge-intensive resources to students, researchers and Qatar's community. Its primary task is to provide access to knowledge aiming to serve research and development (Lux, 2014). Qatar Science and Technology Park was established to become a magnet for companies and institutes from all over the world that needed a place for the development and commercializing of their technologies. QSTP also welcomes entrepreneurs who are planning to start new technology businesses (QSTP, 2016a). Companies that are located there, are expected to conduct commercially oriented research and development while working closely together with the academic institutions and other researchers located in Qatar's Education City (Knight, 2011). Sidra Medical and Research Center was not only designed and constructed to provide world-class patient care for women and children, but also to become a part of the education and research environment that has gained much importance in Qatar (Sidra, 2016). Sidra's patients are nursed in an ultra-modern building that promotes healing (QF, 2015a). Qatar National Research Fund was established in 2015. This Organization aims at funding original, competitively selected research in the academic, public and private sector. It hopes to improve the education and health systems, encourages and quickens innovation and the adoption of new technologies and it aims at a sustainable society and at creating intellectual property. Thereby, it provides chances for many researchers in Qatar ranging from students to professionals (QF, 2015a; QNRF, 2016).

12.1 Libraries and Librarians

"The profession of librarianship in the State of Qatar is currently in a phase of rapid growth, development and transition" (Johnston & Williams, 2015, p. 86). For a long time, librarians in Qatar did not play a significant role in the country. However, with the establishment of Qatar National Library and the will of Qatar to develop a knowledge-based society this is slowly changing for the better. Librarians provide access to the most important resource in this form of society, namely knowledge. The role of librarians in the state of Qatar is now to strongly support education and encourage information consumption and development (Interview partners 2 & 12). At Weill Cornell Medical College in Qatar, "librarians become 'knowledge navigators and information mediators'" (Küster, 2010, p. 18). Moreover, the need for well-trained librarians increases strikingly because they require the skills to "serve multiple groups of clients with distinct library and information needs" (Lux, 2014, p. 177).

Our interviewees are of the opinion that librarians in Qatar are mostly well-trained. The librarians from the university libraries located in Education City sometimes even join the classes at their universities to share their knowledge with the students (Interview partners 1, 3, 8, & 9). Interview partner 2 says that most of the librarians are educated properly to do their job in the country. However, they are trained differently than their colleagues in Western societies. Furthermore, many of the librarians in Qatar only hold an undergraduate degree rather than a master's degree. This is due to the replacement of the library science postgraduate program of Qatar University by an undergraduate LIS program (Daniel, Meho, & Moran, 2015). To change this situation in preparation for the increasing need of well-educated library professionals in the country, the University College London in Qatar offers a master's program in Library and Information Studies (LIS) since October 2013 (Johnston & Williams, 2015). Except for the undergraduate LIS program from Qatar University, which was abandoned in 2012, the postgraduate LIS program of University College London is the first such program in Qatar (Johnston, Mavodza, & Jirjees, 2015). According to interview partner 12, Qatar likes to hire expertise. "Senior library professionals, from overseas, are the most sought after for their expertise and experience in strategic planning and management capabilities" (Kumaresan & Swrooprani, 2013, p. 1350007–2). The majority of the LIS workforce in Qatar comes from Western and other Arab countries. Only about 10 percent of the librarians are Qatari because the quality and experience of Qatari librarians is scarce (Johnston & Williams, 2015). The country hopes that qualified professionals from other countries are likely to train Qatari librarians. It is also important to note that there is a huge difference in the training of librarians in public and private libraries. Nearly every school in Qatar contains its own library that requires library staff (Interview partner 12). This is in line with Al-Kaabi (2015) who states that it is at least intended that every school in Qatar has its own library. In public school libraries, the staff is not trained at all. However, these schools do not place importance on employing professional librarians and on further education for the staff (Interview partner 12). Most of the school librarians only work part-time while lacking the knowledge and needed skills (Al-Kaabi, 2015). School librarians already have the feeling that they receive even less training than other librarians in the country and that they are not appreciated as learning providers. This might become a problem: While Qatar does not focus on supporting school libraries, most of the librarians in Qatar are of the opinion that there is a strong need for school librarians in the country (Johnston & Williams, 2015). Qatar National Library wants to address this issue by offering special training to school librarians (Lux, 2014). Quite unlike public libraries, private libraries do provide training to the librarians (Interview partner 12). However, most of Qatar's librarians state "that they were not satisfied with the level of professional development opportunities that were available to them in Qatar" (Johnston & Williams, 2015, p. 95). Reasons for this opinion are the absence of development support and the requirement to travel to another country to participate in appropriate training. This shows that librarians in Qatar themselves feel the need to expand their professional knowledge and skills to do their jobs accurately. However, the librarians in Qatar National Library are, according to the library itself, described as "experts in teaching, research skills, literacy, programs, and knowledge management" (QNL, 2016b, 4th paragraph).

Qatar National Library combines a national library, a public library, an academic library and a digital library in one building (Interview partner 12).

The library aims at becoming the knowledge hub of Qatar, seeing itself as an important driving force of Qatar's knowledge-based economy. The library's tasks are the provision of published print and online material, the preservation of the heritage collections of the country, the strong support of science and research in the state of Qatar and the offering of public library services as well as the support of Qatar's library network (QNL, 2016d).

> The National Library function collects and provides access to global knowledge relevant to Qatar and the region. It also collects, preserves, and provides access to heritage content and materials about Qatar and the region. The University and Research Library function supports education and research at all levels by providing printed and digital library materials for students and researchers. Metropolitan Public Library delivers library services and information for everyone to enjoy reading, meet people, and become information literate (QNL, 2016b, 2nd paragraph).

After a delay of several years, Qatar National Library opened its new building in Education City in autumn 2017. The library building was planned by Rem Koolhas, a prestigious Dutch architect. In addition to "classical" library services as the offer of printed books and digital resources, the library includes an auditorium with about 450 seats. Here, performances of the Qatar Philharmonic Orchestra will take place. Additionally, there are so-called "Innovation Stations", providing creative tools as 3D printers, a music studio (with lots of musical instruments), audio editing facilities, and tools for video and photo processing. About 200 professionals had already been hired, but without a location it was hard for them to do their jobs (Interview partners 2 & 12). Qatar National Library is, despite that, quite important for Qatar's changing role. Traditionally, public libraries have not been well resourced and used. Most of the people did not even know that they exist. The aim of Qatar is to change this ignorance with the opening of the National Library. Everyone living in Qatar will be enabled to access all of the print and online resources of the library. QNL has purchased national licenses so that the people will be able to use online databases for the acquisition of knowledge. The library allows people to access knowledge everywhere, which is an important aspect in a knowledge-based society (Interview partners 2, 4, 10, & 12). The combination of print and online resources and of different collections will enhance the people's experience and usage of the library (Lux, 2014). Residents have free access to online resources of Qatar National Library that include fiction, non-fiction and research materials (Khatri, 2016). There are already more than 7000 residents registered at Qatar National Library to use the available databases (Lux, 2014). Furthermore, QNL is a major player in library education because it trains librarians and helps them to gain all the knowledge needed to study afterwards at a university and gain their degree (Interview partners 1 & 4). According to Lux (2014), library education is even the main mission of the National Library. Since Qatar National Library is located in Education City, there was once the idea for it to be the main library for students of Education City. The idea resulted from the fact that "Education City currently lacks a central library" (Gyeszly, 2010, p. 87). All the students would use one and the same library instead of visiting the libraries of the branch universities they are enrolled in (Interview partner 12). This project, however, has not been expanded and the building that should have become Education City's central library is now the new establishment for Qatar National Library. And while students might use and benefit from the National Library, they still have to be given the chance to obtain curriculum-focused resources of the branch campus libraries that are specialized on the topics taught by the universities (Flanagan & Wiebe, 2015). According to interviewee 2, the branch campus libraries will cooperate with the National Library when it will be established, but at present the agreement with their home campus is the main source of Education City's university libraries.

When it comes to information dissemination via text and the providing of information in libraries, there is one big problem in Qatar, namely censorship. Books ordered by libraries in Qatar are reviewed before they are imported. If a book argues unlike or against the holy Qur'an, it is not acceptable to list the book in the libraries. Some of the books that are needed have not passed the review. Additionally, some librarians hold back some of the books that arrive at the libraries because they are unsure about putting them on the shelves for the public to read. Sometimes it takes up to two months until a review is completed (Interview partners 1, 2, & 4). Even websites are blocked by Qatar, more precisely by Qtel (nowadays Ooredo) in accordance with the guidelines that are defined by law-enforcing authorities (OpenNet Initiative, 2009), but most of the people living in the region know how to use a virtual private network (VPN). And people in Qatar obviously use it if they need anything they cannot access otherwise (Interview partners 2 & 3).

The libraries in Qatar, especially if they are public, are barely visited. Al-Kutub Library, the former National Library of Qatar, and its branches are rarely used (Lux, 2014). For academic libraries, this fact does not apply because teachers tell their students to go there (Interview partner 12). In general, people in Qatar do not read because "reading and writing are activities restricted exclusively to school" (Ulmer, 2009, p. 112). Our interview partners say that when talking about reading and visiting libraries in Qatar, one needs to distinguish between the habits of Qataris and that of expats. Due to Qatar's nomadic history, a book culture does not exist in the country (Interview partners 1 & 11). "The book is something that is far, far away" (Interview partner 9). A survey conducted by the Childhood Cultural Center of Qatar revealed that children in Qatar are not interested in reading books, stories and magazines (Fahmy, 2015). Expats read and use libraries more frequently than Qataris, but expats that are born in the country probably show similar habits to Qataris because they adapt to the local culture. Maybe, if the needs of Qataris were acknowledged by employing Arabic speaking librarians in libraries, the natives would change their habits regarding visiting libraries (Johnston & Williams, 2015). Just as people in Qatar do not use libraries, they do not visit bookstores. Actually, only 10% of the goods in bookshops are books. Customers of those stores are rather expats than Qataris, but the variety of sophisticated literature is poor (Interview partners 4 & 11).

As always, there are some exceptions. Not everyone in Qatar hates reading, some Qataris actually really like to read and the last book fair was even crowded (Interview partners 1, 2, & 3). The Doha International Book Fair is an annual event where publishers present their books, regardless of genre, to the Qatari community. This event is a chance for Qatar to convince the people of the importance to acknowledge libraries and educational institutions (QF, 2015b). Interviewee 11 says that people in Qatar just skip the step of reading a book. They go from a culture where knowledge is transferred orally to one where knowledge is acquired online. If people are looking for knowledge of any kind, they consult the internet (Interview partner 8). Especially students are required to read, but they prefer to do it online. This holds especially true for Education City's students. Since they are computer literate, they tend to neglect print books or journals and use electronic resources instead (Gyeszly, 2010) whereat they are most interested in articles from social media platforms (Interview partner 1). "To get more people in Qatar to embrace reading, new efforts are underway to make books more accessible to the public, especially children" (Khatri, 2016, 1st paragraph). With the establishment of Qatar National Library and the idea to therein include a Children's Library (QNL, 2016b) with children being recognized as the "core user group" (Johnston & Williams, 2015, p. 92) of libraries, Qatar tries to create a culture that is keen on reading and where using libraries for knowledge consumption is commonplace (Interview partner 12). Interview partner 1 fears that due to their attitude towards libraries, Qataris will not even visit the National Library. Interviewee 12 has also doubts that the library will be well visited. According to him, Qatar National Library is a public library at the wrong place. The location in Education City might be problematic due to its distance to the city center of Doha. Qataris, who do not use libraries anyway, would not go all the way to Education City. Qatar has also recognized this problem and has already thought about opening further branches of Qatar National Library. So far, to attract many visitors, social and recreational aspects, like a cafe and a special events area, as well as cutting-edge technology are included in the library (QNL, 2016a). Moreover, the technical service department and librarians from the public service department promote the new library at public places where many people are reached (Lux, 2014).

Media and information literacy (MIL) is a fairly new aspect in the country, but the Doha Center for Media and Freedom (DCMF) realizes its significance with information becoming more and more important in modern societies. DCMF, a non-profit organization that fights for press freedom and quality journalism, has made it its business to provide media and information literacy education. The organization feels that the younger generations need to be encouraged to become media literate critical thinkers. Since 2011, the organization is actively pursuing and expanding its long-term project to provide MIL education in various schools in Qatar and the region (DCMF, 2013). Information literacy does not seem to be a big topic in Qatar's Education City university libraries given that our interviewees do not know a lot about it. Interviewees 1 and 12 say that information literacy in Qatar is not high. There are some information literacy courses for students in the libraries of the branch campuses, but they call it differently (Interview partners 2 & 3). For instance, the library of the Virginia Commonwealth University in Qatar offers information, media and materials literacy instruction that is tuned to the core curriculum of the University College (VCU, 2014). According to CMU (2015), librarians from Carnegie Mellon University in Qatar hold information literacy workshops in the course of the years. Moreover, if librarians were asked to help, they would, but nobody asks them (Interview partner 1). Qatari librarians feel the need of information literacy instruction. Librarians that teach information literacy are important for the country and information literacy training for librarians is urgently required (Johnston & Williams, 2015). Qatar National Library is aware of the importance of information literacy as well. This field of competence is already among the numerous academic programs and services that the library plans to offer to Qatar's citizens and residents as soon as it starts to operate (QNL, 2016c). Additionally, it plans to offer educational information literacy activities especially for teens (QNL, 2016e). To establish those courses, the National Library aims at working with higher education institutions to make students information literate (Lux, 2014).

12.2 Science and Technology Parks and Science Funding

Just like Qatar National Library, Qatar Science and Technology Park and Sidra Medical and Research Center are part of the large area that forms Education City (QSTP, 2016a; Sidra, 2016). Our interviewees think of QSTP as a good project in Qatar (Interview partners 9 & 10). As a contributor to the National Vision of Qatar, it focuses on the four topics "energy", "environment", "health science" and "information and communication technologies" (QSTP, 2016a). It aims at speeding up the creation of technology in Qatar while strengthening and expanding the support that it offers its tenants (QSTP, 2016e). The designed tasks of QSTP are "to promote academic industry relations, foster innovation and facilitate the transfer of technology from research to the marketplace" (QSTP, 2016c, 1st paragraph). Qatar Science and Technology Park offers world-class facilities where offices are adjusted to high-tech companies (QSTP, 2016b). Many global organizations "like commercial giants such as ExxonMobil and European Aeronautic Defense and Space Company (EADS)" (Alraouf, 2011, p. 59) are already located there and operate from this position. Further innovators get the chance to pursue new ideas and start their own businesses there (Interview partner 1). Even for students, the Science and Technology Park grants facilities. They can do internships in companies located there, like for instance Vodafone, or work for these companies in some other way. The location of QSTP and its closeness to the top universities of Education City is especially beneficial when it comes to research collaboration with these institutions because they foster innovation and creativity (Alraouf, 2011). The research that is conducted at Qatar Science and Technology Park is application-oriented (Interview partner 10). This is one of the criteria that QSTP demands from its tenants to be granted as a member of the Science and Technology Park (QSTP, 2016d). Despite the rather positive point of view of our interviewees towards Qatar Science and Technology Park, so far there is no need for further parks (Interview partner 1).

The attitude towards Sidra Medical and Research Center does not seem to be as affirmative. Just like Qatar National Library, the building has not been completed at the scheduled time (Interview partners 2 & 9). The first outpatient clinic, offering services for women and children, was finally opened in May 2016. The tasks defined in the mission of Sidra Medical and Research Center are the provision of world-class healthcare to its patients, the training of medical students, but also of highly skilled clinicians and the conduction of clinical and transnational biomedical research that focuses on the needs of Qatar's residents and of people from all over the world because it wants to have a global impact. The hospital is an ultra-modern academic medical center equipped with the most secure and reliable technology systems existing (Sidra, 2016; Fig. 12.1). Sidra is a specialized gynecologic hospital which encourages interviewee 4 to raise the question if this hospital is really needed in Qatar. According to him, Sidra Medical and Research Center is starting to close down without ever being opened. Around 200 experts who have been brought to the country to work at Sidra were dismissed before they could even practice their skills. Only a small amount of about 20 scientists is left to constitute the research department. However, with the country's strong focus on the family, which is seen as the main pillar of Qatar's society (GSDP, 2008), there seems to be a need for this institution because Sidra contributes to the nurturance and protection of families in Qatar (Sidra, 2016).

Fig. 12.1 Sidra Medical and Research Center with covered sculptures in front of the building

The Qatar National Research Fund tries to promote Qatar's STM. Adhering straight to the National Vision of Qatar, which is mainly concerned with technology, QNRF aims at encouraging technical development. Research funding is high on the agenda. Every branch of science, but especially engineering and medicine are the fields that receive the most funds. QNRF is not interested in basic research, but in applied sciences because applied science results in outcomes that are patentable. Furthermore, Qatar National Research Fund funds research centers, for instance, for undergraduate programs or conferences (Interview partners 1, 4, 8, & 9). The tasks of QNRF are, beside research funding, the encouragement of dialogue and of partnerships among researchers and their institutions (QNRF, 2016).

12.3 Triple Helix Partnerships

University—government—business partnerships are very rare in Qatar. They do not really exist between all three of these institutions, but rather only between two of them (Interview partners 1, 8, & 9). Interviewee 13 believes that in Education City there are some kind of triple helix partnerships between universities, industry organization and governmental institutions. For instance, the University College London in Qatar works with the governmental institution Qatar Foundation and Qatar Museum on the business side. According to Abduljawad (2015), Qatar tries to stimulate and deepen triple helix partnerships where the government coordinates the relationships between academia and industry. His research revealed that in the state of Qatar partnerships exist between Qatar University, organizations of the oil and gas industry and government institutions like the Ministry of Justice, Qatar Foundation and Qatar Science and Technology Park. Qatar University is linked to the oil and gas industry through the flow of knowledge in both directions. This linkage bases on three pillars: education and training, professional services as well as research and development. While Qatar University is more eager to engage in this relationship, the organizations of the oil and gas industry are doubtful about the precise advantages of knowledge exchange with academia. The governmental institutions are rather passive about this uncertainty of the industry and do not encourage these organizations to collaborate more actively with Qatar University. Not only Qatar University is engaged in triple helix partnerships, Carnegie Mellon University in Qatar also tries to build "strategic partnerships with government and non-governmental organizations across multiple sectors" (CMU, 2015, p. 48; Fig. 12.2). These partnerships, which are just developing in Qatar, have not been formally implemented, supported, or monitored in the country, yet (Abduljawad, 2015).

Even public-private partnerships are not easy in Qatar. There are government entities which establish companies that compete with the private sector instead of working together (Ibrahim & Harrigan, 2012, p. 19). And there are huge problems concerning foreign private companies due to the sponsorship system.

Fig. 12.2 Entrance hall of Carnegie Mellon University with their call to create, inform and connect

12.4 Summarizing Qatar's Further Knowledge-Intensive Institutions

Qatar has recognized the importance of libraries and of research facilities like science and technology parks to pave the way for a knowledge society. With the establishment of Qatar National Library, Qatar Science and Technology Park and Sidra Medical and Research Center the country provides high-quality infrastructure in the fields of education and healthcare and for research and development. The excellently equipped institutions are fundamental to adequately advance these fields from within the state of Qatar. The buildings are not short of the newest technologies and other attractive facilities. Due to this, they have great potential to allure users or experts who attend to the needs of Qatar. Especially the National Library might have a hard time to attract visitors because, due to their culture, people neither read nor visit libraries. However, the country has recognized this problem and tries to encourage reading among the citizens. The developments of the knowledge-intensive institutions are still in an early phase. There is a lack of qualified workers from the country itself. Because of that, experts who are employed by Qatar National Library or Sidra Medical and Research Center are mainly brought in from other countries. Most of them are not able to take up their new jobs because the buildings are partly uncompleted and have not started to operate, yet. The goal of the country is to have expats educate the local younger generations to create high-skilled information literate workers. To really cater for the needs of Qatar, the state has to conduct its own research and development activities. With the establishment of the Qatar National Research Fund, researchers who are concerned with the relevant fields receive support in the form of funds. Moreover, QNRF encourages cooperation of different people and institutions. The close proximity of these organizations to each other and to further knowledge-intensive institutions is therefore of advantage. Especially triple helix partnerships have lately gained attention and Qatar has also recognized the benefit of these partnerships. Some of these linkages do exist in the country. However, the government has not encouraged these partnerships properly, yet, so that academia and especially industry are wary about the sharing of their own knowledge.

As this chapter closes, the detailed analyses of the different aspects regarding the knowledge society and knowledge-based development in Qatar are completed. We have gained many impressions about Qatar's way of transforming its society. It has become clear that the political elites have recognized the need to be supportive and that they intensely try to implement their plans. However, despite the major efforts in planning, funding and constructing, it is also evident that the country still has a long way to go in order to realize all of their appointed goals. What are Qatar's strengths and weaknesses? And which opportunities and threats result thereby? What is the status of Qatar's knowledge-based development and knowledge society? The following and at the same time last chapter is the last piece in the puzzle to answer the all-important question: How does Qatar's future in the knowledge era look like?

References

Abduljawad, H. (2015). Challenges in cultivating knowledge in university-industry-government partnerships: Qatar as a case study. *The Muslim World, 105*(1), 58–77.

Al-Kaabi, M. A. (2015). *Towards activation school libraries in Qatar*. Doha, QA: University College London.

Alraouf, A. A. (2011). Assembling identity for an emerging urbanity: Doha, Qatar as a creative city. In *Proceedings of The City: 2nd International Conference - Culture, Society, and Technology. Vancouver, British Columbia, Canada, 4–6 May 2011* (pp. 48–63).

CMU. (2015). *Carnegie Mellon University Qatar: Annual report 2014–2015*. Doha, QA: Carnegie Mellon University.

Daniel, E. H., Meho, L. I., & Moran, B. B. (2015). Education for library and information science in the Arab states. In A. B. Click, S. Ahmed, J. Hill, & J. D. Marin III (Eds.), *Library and information science in the Middle East and North Africa* (pp. 173–234). Berlin, DE: De Gruyter.

DCMF. (2013). *Supporting media & information literacy education in the Middle East: Experts meeting, Doha, June 11–13, 2013*. Doha, QA: Doha Center for Media Freedom.
Fahmy, H. (2015, September 14). Survey: Qatari youth proud patriots, but lack awareness of other cultures. *Doha News*. Retrieved from www.dohanews.co.
Flanagan, D., & Wiebe, F. (2015). American-style academic libraries in the Gulf region. In A. B. Click, S. Ahmed, J. Hill, & J. D. Martin III (Eds.), *Library and information science in the Middle East and North Africa* (pp. 87–112). Berlin, DE: De Gruyter.
GSDP. (2008). *Qatar national vision 2030*. Doha, QA: General Secretariat for Development Planning.
Gyeszly, S. D. (2010). Qatar's education city's university libraries: Patrons, collections, and services. *Collection Building, 29*(3), 84–90.
Ibrahim, I., & Harrigan, F. (2012). Qatar's economy: Past, present and future. *QScience Connect 2012, 9* (24 pp.).
Johnston, N., Mavodza, J., & Jirjees, J. M. (2015). Exploring the extent to which English LIS Masters programs in the UAE and Qatar meet the skills and knowledge needs of employers in the library and information studies field. In *IFLA WLIC 2015—Cape Town, South Africa in Session 169—Quality Assurance of Library and Information Science (LIS) Education Programs—Education and Training with LIS Education in Developing Countries SIG and Library Theory and Research.*
Johnston, N., & Williams, R. (2015). Skills and knowledge needs assessment of current and future library professionals in the state of Qatar. *Library Management, 36*(1/2), 86–98.
Khatri, S. S. (2016, April 11). Qatar rolls out national reading campaign. *Doha News*. Retrieved from www.dohanews.co.
Knight, J. (2011). Education hubs: A fad, a brand, an innovation? *Journal of Studies in International Education, 15*(3), 221–240.
Kumaresan, S. C., & Swrooprani, B. S. (2013). Knowledge sharing and factors influencing sharing in libraries: A pilot study on the knowledge sharing attributes of the Education City library community in Qatar. *Journal of Information & Knowledge Management, 12*(1), 1350007 (13 pages).
Küster, A. (2010). Qatar—a new beginning? The development of a collaborative research and education environment in education city, Doha. *Serials, 23*(1), 16–19.
Lux, C. (2014). Qatar national library: Architecture as innovation in the Arab world. *International Federation of Library Associations and Institutions, 40*(3), 174–181.
Mainka, A., & Khveshchanka, S. (2012). Digital libraries as knowledge hubs in informational cities. In *Libraries in the Digital Age (LIDA) Proceedings Vol. 12, 18–22 June 2012*. Zadar, HR: University of Zadar.
OpenNet Initiative. (2009). *Internet filtering in Qatar*. Retrieved from www.opennet.net.
QF. (2015a). *Annual report 2013/14*. Doha, QA: Qatar Foundation.
QF. (2015b). QNL participates in 26th Doha International Book Fair. *The Foundation, 84*, 8.
QNL. (2016a). *A unique library with breadth in its functions, bridging with knowledge Qatar's and the Arab and Islamic World's heritage and future*. Doha, QA: Qatar National Library.
QNL. (2016b). *About the library*. Retrieved from www.qnl.qa.
QNL. (2016c). *Academic programs and services*. Retrieved from www.qnl.qa.
QNL. (2016d). *Director's welcome*. Retrieved from www.qnl.qa.
QNL. (2016e). *Educational programs and services*. Retrieved from www.qnl.qa.
QNRF. (2016). *Vision & mission*. Retrieved from www.qnrf.org.
QSTP. (2016a). *About us*. Retrieved from www.qstp.org.qa.
QSTP. (2016b). *Facilities and services*. Retrieved from www.qstp.org.qa.
QSTP. (2016c). *Free zone*. Retrieved from www.qstp.org.qa.
QSTP. (2016d). *Requirements*. Retrieved from www.qstp.org.qa.
QSTP. (2016e). *Vision & mission*. Retrieved from www.qstp.org.qa.

Sidra. (2016). *Sidra medical and research center*. Retrieved from www.sidra.org.

Stock, W. G. (2015). Informational urbanism. *Journal of Systemics, Cybernetics and Informatics, 13*(6), 62–69.

Ulmer, J. (2009). The knowledge society: Text and orality in Qatar. *The International Journal of the Book, 6*(1), 109–113.

VCU. (2014). *VCUQatar information, media, and materials literacy program: Collbaorative learning outcomes*. Doha, QA: Virginia Commonwealth University.

Part IV
Conclusion

Chapter 13
Strengths, Weaknesses, Opportunities and Threats of Qatar's Way into the Knowledge Society

Abstract Qatar is a country with diverse strengths that lead to opportunities when it comes to the construction of a knowledge society. However, the country also has to deal with several weaknesses and threats, which might stand in the way of a positive development. It is on Qatar whether it uses the circumstances by which it is surrounded to strengthen the country and successfully master the transition or not. The SWOT-analysis reveals four possible scenarios about Qatar's future.

Keywords Qatar · Knowledge society · SWOT analysis · Strengths Weaknesses · Opportunities · Threats · Knowledge · Future scenarios

Qatar has realized the need to change its ways in order to become a knowledge society and has, therefore, come up with many ideas and plans paving the way for this major transformation. The due date to which the state wants to have achieved the goal of being a fully established knowledge society is the year 2030 and it has already invested in and implemented many projects resulting, along with some other advantages, in some important strengths. However, not everything has gone as planned and some obstacles still need to be overcome to fully succeed. Out of these strengths and weaknesses, there arise some opportunities and threats that the country might encounter while trying to become a renowned knowledge society recognized by people worldwide.

13.1 Strengths

The probably most fundamental strength of the state of Qatar is its strong will to develop a society based on knowledge. All plans—regardless of depending on the country's economy, like the diversification of the industries, or on other things, like the education of Qatari and expatriate individuals—are designed for "Qatar's aim to become a premier knowledge-based society" (GSDP, 2011, p. 122).

J. Gremm et al., *Transitioning Towards a Knowledge Society*,
https://doi.org/10.1007/978-3-319-71195-9_13

The investments of the country take one specific path: the development of knowledge-intensive institutions and facilities. The prestigious branch campus project, the construction of the new National Library, Sidra and the Science and Technology Park are just some examples. International universities provide a high-class study for the young Qatari and expatriate population and therewith bring the education to the students instead of bringing the students to the education as it is the case in most parts of the world. The National Library serves as a knowledge base offering easy access to knowledge to everyone. Sidra shall not only be a good medical institution, but rather a scientifically recognized research institute. The Science and Technology Park houses a large number of technology-based companies providing internship places for Qatar's students and graduates and, in addition to it, supports the development of start-ups.

However, besides investment in buildings and the people who work there—almost entirely engaged from outside the country—the government supports the upgrade of research. By importing knowledge in the form of expats that are experts in their field, providing excellent research opportunities and funds and thus getting output for the development of new economic ways and branches, the country tries to create an "economy [that] diversifies from its reliance on gas and oil" (GSDP, 2011, p. 13). Due to the fact that the hydrocarbon reserves run out steadily, it sees an expedient in generating previously undiscovered sectors. The diversification process in implementing a MICE industry in Qatar was quite successful. The Qatar National Convention Center and the Doha Exhibition Center are nowadays well-known locations for international conferences and periodically attract scientists and business people.

The large investments in knowledge additionally result in a further key advantage for generations to come. The multi-layered university landscape composed of the universities in Education City and Qatar University offers first class study opportunities. That way, the mainly young population that is in itself an additional strength of Qatar, has manifold opportunities to upgrade its education (Interview partner 5). The level of education and the quality of research are comparable to Western standards. Teachers and students are equally urged to do research, to develop new ideas and to run different projects that will have a notable effect on the country, its economy and its society. Educational activities in the country and especially the Education City project are supported and brought into being by Sheikha Moza, one of the founders of Qatar Foundation for Education, Science and Community Development. Furthermore, the country and its leaders ensure the implementation of any project by a generous research fund that is comparable to those of the United States, Germany or Denmark. With the implementation of the branch campuses, also women, who are the most hardworking students and build the most promising future group of workers, have the opportunity to study in the country—in contrast to previous times even in the disciplines of engineering and science. The fact that most of these women are not allowed by their families to leave the country alone, would have ruined the chance of receiving Western education. To simplify the young generation's start into the student life, Qatar is changing the outdated K-12 system—even consults professionals to improve the rather unstructured old system—so that they have no longer any problems when entering one of Qatar's universities.

In turn, the variety of educational institutions and offers causes an advantage for the country's economy, namely, that all existing branches will be well served by graduates. The distinctive cooperation between both parties—the universities and the local companies—provides this transition into the labor market with the opportunity of a good preparation of the students by internships for a successful integration and motivation boost. Despite the traditional way of thinking, Qatari companies do not differ between men and women when choosing new employees.

Owing to the internationalization within the universities' research and study teams as well as within the whole region, different cultures and attitudes from all over the world result in a promising variety of ideas—an infrastructure of research arose and collaborations all over the world were made available (Interview partner 9). Due to this fruitful, intercultural cooperation, we can say that toleration is existing in the country.

All those aspects like the import of experts or the invitation of prestigious universities could just be realized due to "the availability of capital, which allows cost intensive infrastructure investments and the ability to launch various new knowledge initiatives" (Conventz, Thierstein, Wiedmann, & Salama, 2015, p. 17). The accumulated wealth of the country enables to realize probably every plan stated in the National Vision—at least those that can be realized through investments. The economic potential of Qatar indeed goes along with another key dimension: "[T]he geographical position of Doha within the Gulf region itself as well as markets in Asia and Europe" (Conventz, Thierstein, Wiedmann, & Salama, 2015, p. 17) enables the country to act globally.

13.2 Weaknesses

However, there is no profit in the wealth, which predominates in the country, if the overall citizens' behavior does not fit the intended purpose publicized in plans and visions. Since the country still banks to a great extent on the oil and gas reserves there is no real motivation for the Qataris themselves to change the economy. They rather remain seated on their positions in the public administrative sector where they are paid well for less work, or occupy a—due to the Qatarization program and not due to their qualification—new found working place in the private sector, instead of working with motivation and efficiently in one of Qatar's universities or a private company or even start a business themselves, for example, in the knowledge sector. The children's wealth—especially that of the boys—is too large and they are not thinking critically about this—Qataris are not hungry for knowledge and success, but rather they are satisfied with their life (Interview partner 10). The atmosphere is not exactly innovative which leads to the import of more and more experts, who indeed secure this lifestyle, but likewise threaten the traditional, non-Western way of behavior.

Some development plans fail, probably because the country tries to implement its purposes within a too short time period. Promising projects like the launch of a new education system—whose implementation is desperate due to the insufficient previous system—simply take time and are not realizable within the short time period the government wishes for. "The lack of maturation of Qatar's educational system, lack of education-workforce alignment, segmented labor market (nationals in government jobs, expatriates in the private sector), and problems associated with the burdensome and inefficient public sector have been known for at least two decades", Weber (2014, p. 79) concludes.

This is also evident considering that several construction projects like Sidra or the National Library have not been finished within the scheduled time period. In addition, the rather low number of patents and publications plays a part in contributing to the dragging diversification of the economy. Without any ideas, there are no patent applications and then again no new businesses. The research quantity is definitely too low and is nowhere near Western outputs. Furthermore, for teachers it is hard to perform teaching and research at the same time because the students need a lot of support, which takes time and keeps them from performing research. Since cooperation between the scientists and the political elites does not exist, there is no exchange of profound knowledge about new trends, technologies and development opportunities. Especially in the information industry, there is too little progress. Knowledge-intensive companies and start-ups are rather rare.

However, the country is probably in disagreement with itself about where its path should lead to (Interview partner 13). Reasons for the large non-participation in the country's programs could be the imperfectly expanded government-citizen relationship. When regarding the social media activity, the communication is mainly one-sided, from the top down—from the government to the citizen. Indeed, the citizens react to the government's postings, but do not give voice to their thoughts and needs by posting content on the remaining governmental social media sides themselves. Without clear utterances, the government cannot do much and makes decisions without including anyone else in the decision-making process. Therewith, it probably snubs the nation that in turn gets even more demotivated to change the existing situation, which then again affects the government negatively. This leads the country to a dilemma: The government does its own thing, makes all economic and country-specific decisions without getting the citizens' opinion and the Qataris do not come up with the idea of breaking out of their unambitious, luxury life by developing new concepts and motivation. Due to the lack of motivation, Qataris are still depending on the expats, their knowledge and their working morale. They did not find a way to collect and keep the knowledge in their country until now. So if the expats leave, they will take the knowledge with them (Interview partner 10). As a consequence of this reliance and the will to keep the expats in the country, the Qatari culture gets more and more lost in a state that is very international and full of foreign talents. With English and not any longer Arabic as the most spoken language in the city region, Qataris retire to the desert and settle down there in little camps for some time. They are tolerant and open just due to the fact that they do not have another option. In fact, they are rather conservative and restricted (Interview partner 1).

However, this way of life—being free because of wealth, but depending on those who work for this wealth—involves a danger. The presence of the *kafāla* system results in the circumstance that even now some expatriate workers leave the country because of missing sponsorships for themselves or their children. The jobs are in general unsecure (Interview partner 9). For the expats, it is sometimes a hard decision, because meanwhile they started to forge an identity in Qatar. For others, it states no problem because they will leave the company or the country anyway if somebody else pays more or offers better conditions (Interview partner 8). Others never lost their connection to their home country—sometimes due to their relatives, sometimes due to the rather bad quality of physicians in Qatar and their regular stays in their home country to get adequate medical help. In contrast, Qataris have to fear that they lose their workers. Indeed, this is probably not a problem because they will always find replacement. Actually, the most promising future Qatari working group consists mainly of women. They dominate the universities and put the males—who rather leave the country to study abroad so that the good Qatari student body is small—therewith in the rear. Despite the women's motivation, they are restricted in their opportunities. Most of them are not allowed to go outside the country and study at another university. Even so, with the implementation of the branch campus project Education City, they are offered good, wide-ranging conditions for studying in their own country. However, after their marriage a glorious career is in many cases not in sight—which depends on the husband—and the realization of the women's ambitious schemes is interfered. In addition, it is uncommon that Qatari women accomplish to get a leadership position. This traditional way of thinking and acting is also reflected in the fact that, for instance, courses at the public university are still gender-segregated. In some other cases, women only decide to study to increase their reputation and not to make a career.

A further weakness of Qatar arises when regarding the Arabian culture, but the perspective is different. From the traditional perspective, reading—that is important to develop new knowledge, new ideas and therewith new businesses—is not prevalent in Qatar even though it is an aspect of the culture. The Qatari culture of reading is something that still has to emerge. Furthermore, the librarians who could give the Qataris an understanding of (scientific) reading, are not well trained and the access to certain information is denied due to the fact that some books or their content are not accepted by the holy Qur'an.

Taken as a whole, Qatar is multiply divided. On the one hand, there are two groups of Qataris—those who try to be more open and accept new challenges and those who are highly conservative and do not believe in striking new paths, which might be against the culture or the religion. On the other hand, Qataris have different opinions in general and in regard of expats—that are again divided into talents and low-skilled workers. To pave Qatar's way into the knowledge society will be a successful project, only if all these groups cooperate. Otherwise the country will fail because of clashing opinions.

13.3 Opportunities

As strange as it sounds, the depletion of natural resources or the decreasing demand of the oil-burning countries due to the increase in environmental awareness and use of electromobility might be one of the biggest opportunities of all, given that Qatar has no other option than changing its society. "Qatar has to be needy" (Interview partner 4)—it has to understand what it means to really need something. Due to its prosperity, the country has not experienced the absence of essentials in the last decades. Only if Qatar really recognizes the meaning of the threat of the decreasing oil and gas reserves, the eagerness to diversify its economy and to implement knowledge-based development might be taken even more seriously by Qatar's entire population. The arising opportunity is also supported by the fact that Qatar has made the decision to change and expand many aspects of the society. The government has already developed plans to enforce these changes (Interview partners 2 & 13). Due to this positive attitude of the political elites, the country has the opportunity to follow through with all of its appointed tasks aiming to master the transition into the knowledge society.

As the country has realized that it is in need of large numbers of knowledge workers to achieve the desired development and that there are not enough talented Qataris to fill the positions, it has started to attract foreign talents to work in the local companies and institutions. Only with the support of these experts and their knowledge, Qatar has the opportunity to successfully pursue its plans to diversify its economy and to establish a knowledge society. On the one hand, they bring knowledge into the country and, on the other hand, they train the locals and thereby share their knowledge with them, preparing them for knowledge-intensive jobs.

The country knows that it is urgently required to bring locals into the knowledge-intensive positions in the private sector. As a measure to accelerate this process, the country introduced the concept of Qatarization. If this plan is implemented properly in the future and Qataris are preferred to expats not only because of their origin, but also because they are able to compete with them, this concept might be an opportunity for Qatar to reduce or eventually even put an end to its reliability on foreign knowledge workers.

However, while Qatarization is one possibility for the country to become independent, some of Qatar's further strengths provide the same opportunity. Alarmed by the poor performance of local students, Qatar has decided to improve its entire education system. It has undertaken reforms to change its K-12 education system and the direction of Qatar University, which might improve the qualification of the local students, given that those responsible grant enough time to implement these reforms. Additionally, branches of renowned universities from overseas were opened in Qatar and large investments were made to establish a world-class infrastructure for all education institutions, which were stuffed by well-educated experts from abroad. Due to these changes, the country now offers high quality higher education conditions to the younger generations of Qatar, which is a major opportunity for the next generation of leaders and, with that, for the development of the country.

Whether Qataris really use the chance to acquire knowledge and skills in these institutions strongly depends on their parents as their role models. If parents are supportive of education and encourage their children to work hard, there is a chance that the mentality of young Qataris will be defined by ambition in the future.

If the students are eager to acquire knowledge and skills, they will be well prepared for their future jobs in Qatar. "Students who decide to undertake higher education in their home countries are less likely to consider emigration and therefore the 'brain drain' of Gulf nationals is reduced, and local economies can benefit from the knowledge and skills acquired by these students" (Wilkins, 2011, p. 77). The country will profit from this at least by being able to draw on its own knowledge base. In the long run, Qatar will not have to import foreign knowledge anymore.

It has crystallized that females—for whom the opportunities to study have increased in the last few years—study more seriously and more successfully than males in the country. The fervor of the women might therefore be the starting point to bring educated Qatari females into the leading positions of Qatar's companies and administrations. This might also result in changes of the society with regard to gender equality.

Qatar has realized that, beside education, research and development is an important part of society development. The country encourages and supports its citizens in this regard as it tries to establish an innovative atmosphere. The citizens have access to research and development in terms of funds and needed resources, which may lead to an improvement of scientific output. A tiny success is already visible and if the country continues to carry out research and to make inventions, the next generation might become a research generation and Qatar might make the breakthrough needed to compete in the 21st century.

Due to the fact that many well-educated foreigners have come to the country, Qatar has become an international location where different views clash. For Qatar, this might be an opportunity to incorporate these different views in order to attain a good general perspective. For instance, all of these expatriates can potentially contribute to the research and development process by sharing their knowledge and experiences. Outstanding research results would thus bring the country to the fore. Perhaps, the conglomeration of global knowledge might even be an opportunity for Qatar to become a major player in the global economy given that knowledge from all over the world is pooled in the small country.

Qatar has already concluded some good deals with regard to this change of direction. Its plans to diversify the MICE industry and the thereby already achieved goals as well as hosting some of the world's most popular sports events provide the opportunity to attract people, to make money, to raise visibility and to make new contacts (Interview partner 2). This opportunity is supported by Qatar's strategic location between Europe, Africa and Asia, which might be a benefit for the country when it comes to logistics and trade as well as travelers. The country has its own airport and good connections to all continents via Dubai (The Emirates Group, 2016).

Also, the country has made some progress when it comes to its tourism industry. Tourists have different intentions than MICE-participants and businessmen. Only if Qatar manages to keep visitors in the country for more than a couple of days, the tourism industry will generate notable profit.

Making new contacts and establishing long-term partnerships is not only intended on a global scale, but also on a local scale between governmental, corporate and academic institutions within the country. Triple helix partnerships between research institutions, public administrations and companies are in the early stages. These collaborations might help to improve research and to apply the available knowledge in the most efficient way.

The conglomeration of knowledge-intensive institutions contributes to the establishment of partnerships. Education City assures not only the spatial proximity of universities among themselves, but also of the universities to Qatar Science and Technology Park and to the new Qatar National Library. The latter is envisaged to be a preserver of the entirety of Qatar's knowledge and is established to be the first place to go to when knowledge is needed—not only for the establishment of a knowledge society. While the buildup is well underway, the citizens are already encouraged to change their habits and to start reading in order to contribute to the plans of the political elites.

However, the book is not the only medium the country tries to promote. Qatar has recognized the need to improve media and information literacy in Qatar and has started to implement courses in the country's libraries. If this is successful and if these courses are well-visited, the citizens might be able to even more suitably support the knowledge society.

With Qatar Science and Technology Park, the country provides a place for knowledge-intensive companies and start-ups. In general, the country is eager to attract fruitful companies and at the same time supports the establishment of start-ups. This might help to reinforce the local business.

However, Qatar is also a mighty supporter of investments abroad and tries to gain revenues by holding shares in foreign companies. As long as this thrives, this is a lucrative chance for Qatar to make large amounts of money.

Another field where Qatar made enormous investments is healthcare. Ultramodern infrastructure was established for hospitals and the associated research institutions, like Sidra, and qualified foreigners are employed at these institutions. If these institutions fulfill their purpose, the quality of life in Qatar will rise significantly. Maybe even foreign talents who are often suspicious of Qatar's health system and still travel abroad for medical treatment will gain more trust. This will raise the attractiveness of Qatar, will bring more expats to the country, will keep them in the country and will furthermore bring in more money.

The construction of the new railway system supports the visitors of the new knowledge-intensive institutions by providing a fast and easy access to the buildings and, in general, this means of public transport will erase the traffic problems in Qatar once its construction is finished.

13.4 Threats

Just as it is one of the greatest opportunities, either the depletion of Qatar's oil reserves, which most certainly will happen sooner or later, or the decreasing demand, which might happen even sooner, is also the biggest threat. How big of a threat this is for the country depends on the leaders', but also on the people's dedication to realize the scheduled measures for establishing a knowledge society. Right now, the pressure is not high enough to get every resident to participate in the changes—most of them stick to the rentier mentality, which is predominant in the country. Many Qataris do not aim at a successful graduation and continue to apply for jobs in the public sector while leaving the meaningful, knowledge-intensive positions in the private sector companies to be occupied by foreign talents. The most qualified scientists, engineers and physicians are imported from other countries. If there is no change in this state of affairs, then the dependence on the foreign workers and their skills will remain. While this primarily seems to be the decision of the working population, the political leaders are partly to blame for this, because they are in charge of the different working conditions in the public and private sector. As long as these conditions will persist the way they are now, there is an insufficient incentive for Qataris to work for their money. Indeed, with the concept of Qatarization, the authorities have demonstrated their ambition to get nationals to work in the private sector. However, as long as Qataris are not chosen by employers exclusively by their abilities, the really talented workers will not acquire positions where they can achieve something, so that this concept is going to backfire in the long term. Furthermore, the enforcement of Qatarization chases expatriates away, because it becomes much harder for them to find appropriate employment. However, this is not a huge problem for expats, but rather for Qataris, since expats will easily find jobs in other countries while Qataris, in this case, will have to work by themselves and are expected to conduce to success. If they are not willing or able to do so, nobody will diversify the economy.

The motivation of Qataris to study and to become part of the high-skilled workforce strongly depends on the opinion and points of view of their families. If parents do not encourage their children to study hard and to search for a good job afterwards, the children will probably not find that motivation themselves. In order to attend one of the best higher education institutions in the country, Qataris need to find this motivation fairly early, because they need to acquire sufficient knowledge to pass the acceptance test of one of the elite universities. Often they fail due to lacking skills in English, which is the teaching language of all the branch campuses in Education City.

While Qatari women—who study more often and more successfully than males—might become the ones to hold the knowledge-intensive positions, there is the danger of tradition. As the society of Qatar is traditionally male dominated and women have fewer opportunities (such as they are traditionally the ones staying home to care for the family), it might become harder for women to assert themselves. If they cannot, there will be only few Qatari graduates who become part of the qualified workforce in the country and the expats will hold the upper hand.

At the same time, Qatar cannot rely on the willingness of foreign experts to come to the country. Due to the *kafāla* system, expats have too little freedom and guarantee and, although tolerated, they are not treated with the openness they deserve, so that they might not work as good as they could to make the company, and therewith the country, successful. So far the country has had the financial resources to attract people, but cannot retain them for more than five to ten years resulting in the fluctuation of knowledge. So, while Qatar is good at bringing foreigners to the country to work on site, there is simultaneously the threat of foreigners' fading interest in the country, may it be due to reduced payment or unsupportable decisions of the country. Both is possible, since the country has to economize—a measure caused by the decreasing oil prices—it might react by offering less salary to the expatriates. Since other countries are interested in foreign workers as well, there is no reason for them to go to or to stay in Qatar (Interview partner 4). Especially if the country reinforces the effort to switch the language of communication from English back to Arabic, the expats will feel unwanted and consequently search for employment elsewhere. This retreat to traditions shows that Qataris fear the emerging Westernization and thus separate from expatriates more and more, dividing the country even further. The result is that there are too many different opinions when it comes to changes in society and people do not work together, but rather against each other, and the will to support the transformation into the knowledge society might fade.

As the oil continues to bring in less money, all funds that are invested into to the establishment of the knowledge society—not just the ones appointed for expats—will go down. As a consequence, the whole development is on hold (Interview partner 3). Moreover, the support of research and development, which is – given the low numbers of publications and patents—not a very productive and accepted field in the country, yet, will be withdrawn. Without the support, there is no further innovation and consequently, the country does not have any advantages in global competition.

The trimming of investments, however, is not the only reason why some of the plans are canceled or do not bring the anticipated success. Additionally, the country's way to follow through with its plans sometimes gets in the way of a successful result, because it does not give enough time for the proper implementation of reforms. If Qatar continues to force changes as it does now, the projects are predestined to fail.

Another threat arises from the fact that the country makes decisions without regarding the needs of the citizens, as they do not consult scientists or engineers to establish plans on the background of funded knowledge and as the communication between the government and the people is organized as top down information exchange. The political leaders make decisions that cannot be realized, they are unaware of the citizens' opinions and do not know how to motivate them, which might lead to dissatisfaction and further non-participation. These are attributes that should not be present in a country that tries to establish a knowledge society.

The country should change some traditions and encourage its people, especially the younger generations—namely students—to participate. For instance, the idea of studying just to enhance the prestige (Interview partner 4) needs to be combated and the teaching in Arabic at the public education institutions as well as the tradition of gender-segregated studies, which is still commonplace in Qatar's public university, have to be reconsidered in order to prepare students for future conditions at the labor market. The neglecting of the last two would lead to the incapacity of Qataris to work in an international company or in a mixed-gender environment, which would ruin chances for the local companies and thereby harm the whole economy.

Not only the offering of gender separated studies bears this risk, but also the habit that students who are sponsored by local companies seek employment in exactly those companies after their graduation. This gets in the way of the concept of Education City that every branch university serves a different economical sector and furthermore results in the employment of job beginners unqualified for this position. Consequently, Qatar's economy does not take the leap to establish productive non-oil sectors, resulting in the danger of them relying fully on their investments abroad so that they will never be able to stand economically on their own two feet (Interview partner 11). Furthermore, there is always the looming threat of bankruptcy of those companies that Qatar invested in, which cannot truly be influenced by the country. Even though the investments are promising and as a matter of fact involve global companies, Qatar needs to lay the focus on its own knowledge-intensive enterprises. Especially the number of companies in the knowledge sector is insufficient, resulting in the lack of workplaces for knowledge workers—who form the group of workers that is crucial in a knowledge society as it brings in profit.

Even though Qatar has established the new Qatar National Library, it is not certain that this institution will bring the anticipated success. Since people in Qatar are said to be rather uninterested in reading, the Qatar National Library might not be used, especially as it is situated all the way out in Education City and therefore cannot be reached that easily.

The location of the country right between Europe, Africa and Asia is an outstanding possibility for the expansion of its companies to become global players. Even so, Qatar's "firms focus their activities mainly on local markets rather than on using their location for doing business beyond the borders of Qatar" (Conventz, Thierstein, Wiedmann, & Salama, 2015, p. 17). Thereby they miss the chance to become acknowledged, not only for their oil resources, but also for success in other business sectors.

13.5 Hypothetical Scenarios

Based on Qatar's strengths, weaknesses, opportunities and threats, different possibilities about the country's future arise. The most important factors that might affect the development of the country are the depletion of natural resources, the

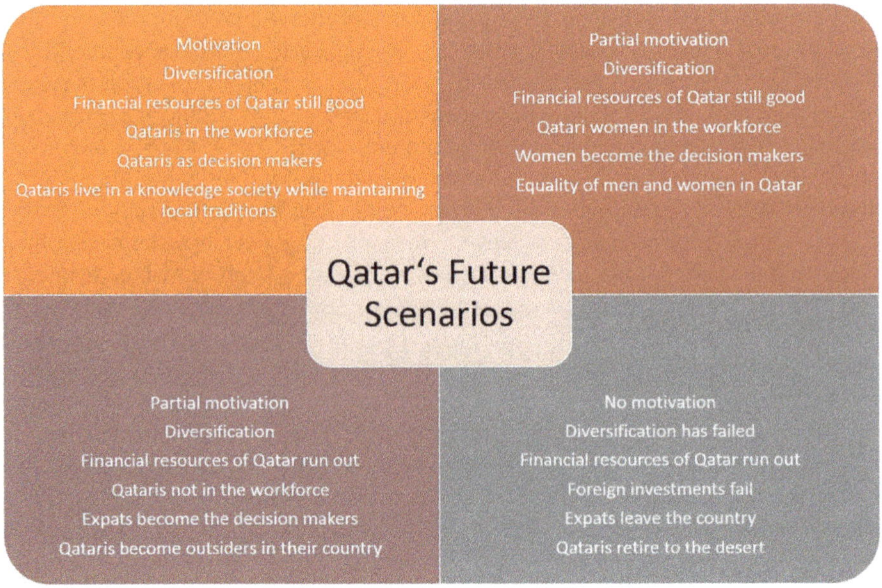

Fig. 13.1 Qatar's future scenarios

diversification of the economy, the country's financial resources as well as the motivation and qualification of the political leaders and Qatar's citizens. As the oil vanishes and with it the primary source of Qatar's prosperity, the drastic change of the country is predetermined, resulting in four hypothetical scenarios (Fig. 13.1).

(Scenario 1) Infected by the determination of the politicians and their general engagement in the establishment of knowledge-based development and of a knowledge society, Qatar's citizens have found new motivation to develop their knowledge and skills and have successfully established a strong and diversified economy. While having been supported by foreign talents in the beginning of this process, they are now able to adequately replace those imported experts and occupy the major positions in the public and especially the private sector. The country can fully rely on its own knowledge base while being globally competitive without depending on investments abroad. All planned projects that were part of the National Vision 2030 are completed and the construction site that Qatar once was, is now a neat space of ultra-modern skyscrapers and charming corniches that is known as a leading center for culture and excellent education, while it attracts many international guests from tourists to MICE participants.

(Scenario 2) The motivation to acquire new knowledge has only been evoked in women. While Qatari men have rested on their financial protection and did not find the motivation to be involved in the change of society, Qatari women have become more and more ambitious and are hence still the ones with better academic degrees—surpassing the men increasingly. This was a critical situation in a male dominated society where women had much less rights and privileges, but the females have asserted themselves by not accepting their traditionally minor positions and are now occupying the leading positions in the private and public sector. Females are now the decision makers not only of the companies in Qatar, but also of the whole state.

(Scenario 3) Qatari males have not found any motivation to really engage in the change of society and the eagerness of the women has either never happened or the women striving for success have been suppressed by men who were determined to maintain the local traditions. So, Qataris have never accomplished to be part of the most privileged workforce and neither Qatari men nor Qatari women will be able to get into these positions. As a consequence, expats still occupy the knowledge-intensive and decisive positions in Qatar. They are well payed by the locals, which motivated them to diversify the whole economy on their own, without great assistance of Qataris. As expats slowly but surely gained the upper hand, they have taken over all of the leading positions in the local companies and sometime later they even occupied the leading positions in the country, so that Qataris—significantly outnumbered by expats—do not have a say in the governance of their own country any longer.

(Scenario 4) The diversification has failed. Qataris lacked and still lack the skills to establish a knowledge society and the financial resources are not sufficient to keep the foreign experts in the country in order to continue helping the country to climb the ladder of success. All of the imported talents have left Qatar to seek employment where the payment and working conditions are better. They have taken all the knowledge and skills with them, leaving the locals to their own fate. As Qatar has earned the majority of its financial means through successful investments abroad, it has lived of these means as long as the companies yielded good returns. However, as some of the companies that Qatar had invested in depreciated, the country now has nothing left to be able to play in the upper league. With the oil and the thereby achieved prosperity long gone, the country is forgotten as fast as it had been built up and brought to life. The ruins, which once had been futuristic buildings—never completely finished—derelict more and more. What is left in the end is a vast everlasting desert that houses a poor population living in lowly tents and leading a traditional nomadic-like life.

Only time will show which future scenario becomes reality.

References

Conventz, S., Thierstein, A., Wiedmann, F., & Salama, A. M. (2015). When the oryx takes off: Doha a new rising knowledge hub in the Gulf-region? *International Journal of Knowledge-Based Development, 6*(1), 65–82.

GSDP. (2011). *Qatar national development strategy 2011–2016*. Doha, QA: General Secreteriat for Development and Planning.

The Emirates Group. (2016). *Annual report 2015–16*. Dubai, AE: The Emirates Group.

Weber, A. S. (2014). Education, development and sustainability in Qatar: A case study of economic and knowledge transformation in the Arabian Gulf. In A. W. Wiseman, N. H. Alromi, & S. Alshumrani (Eds.), *Education for a knowledge society in Arabian Gulf countries* (pp. 59–82). Bingley, UK: Emerald.

Wilkins, S. (2011). Who benefits from foreign universities in the Arab Gulf states? *Australian Universities' Review, 53*(1), 73–83.

Author Index

J. Gremm et al., *Transitioning Towards a Knowledge Society*,
https://doi.org/10.1007/978-3-319-71195-9

Subject Index

J. Gremm et al., *Transitioning Towards a Knowledge Society*,
https://doi.org/10.1007/978-3-319-71195-9

CPSIA information can be obtained
at www.ICGtesting.com
Printed in the USA
LVHW02*1315271217
560944LV00001B/67/P